Nordia
Geographical Publications

Volume 35:2

NGP Yearbook 2006

Theme Issue on Northern Nature and Human Activities

Guest editor Olavi Heikkinen

Editors Katri Suorsa, Marja Anttonen & Marja Mönkkönen

Nordia Geographical Publications

Publications of
The Geographical Society of Northern Finland
and
The Department of Geography, University of Oulu

Address: Department of Geography
P.O. Box 3000
FIN-90014 University of Oulu
FINLAND
marja.monkkonen@oulu.fi

Guest editor: Olavi Heikkinen

Editors: Katri Suorsa, Marja Anttonen & Marja Mönkkönen

Nordia Geographical Publications
ISBN 951-42-8326-0
ISSN 1238-2086

Multiprint Oy
Oulu 2006

CONTENTS

1 Northern Nature and Human Activities - Introduction
Olavi Heikkinen

3 New Applications of Political Ecology for the Nordic Context
Paul Robbins & Hannu Heikkinen

9 The Potentiality of Remote Sensing in Biogeographical Research
Miska Luoto

17 Combining Data from Satellite Images and Reindeer Herders in Arctic Petroleum Development: the Case of Yamal, West Siberia
Timo Kumpula, Bruce Forbes & Florian Stammler

31 The Forgotten Uranium Mine of Paukkajanvaara, North Karelia, Finland
Alfred Colpaert

39 Location-based Services as Tool for Developing Tourism in Marginal Regions
Harri Antikainen, Jarmo Rusanen, Sami Vartiainen, Mauri Myllyaho, Jari Karvonen, Markku Oivo, Jouni Similä & Kari Laine

51 From the World of Geography

Excursion on the Edge of Europe
Marja Anttonen, Tanja Löytynoja & Katri Suorsa

57 List of Contributors

Northern Nature and Human Activities – Introduction

Olavi Heikkinen
Department of Geography, University of Oulu

The north, the northern regions, and the arctic. What they are? Do they have boundaries? All places and phenomena in the world have a northern dimension, a northern location.

The north is a direction. Chicago is north of Mexico City, Oulu is north of Helsinki. The North Pole, which is situated in the middle of the Arctic Ocean, is the ultimate north. Far in the north, for instance in Lapland, the north is almost synonymous with the arctic. The areas surrounding the North Pole receive very little solar radiation, and in winter only indirect solar radiation. For this reason those areas are cold environments with a short growing season and a summer.

The arctic certainly lies to the north, but the scientific methods to define the arctic can be disputed. Further, it depends upon the scientist's own background how to determine what is the arctic.

The following criteria for the arctic have often been presented: (1.) The Arctic Circle, an astronomic boundary, has been suggested as the southern limit for the arctic. There the sun during winter remains below the horizon. (2.) The northern tree-line, primarily controlled by climatic conditions, is also frequently seen as the southern limit of the arctic. The tree-line is easily visible in the landscape. It is an ecologically important ecotone between the boreal forests and the treeless tundra. It is the northern limit for forestry and agriculture. (3.) The 10 °C isotherm for the warmest month of the year is suggested to form the southern boundary of the arctic because the northern tree-line tends to follow this isotherm. (4.) Permafrost characterizes the arctic, although permafrost areas can also be found at relatively southern latitudes, and sometimes the ground can even support forests. This is the case particularly in areas with a continental climate. (5.) Glaciers and sea ice are typical of northern and arctic areas. (6.) Northern lights are a polar and arctic phenomenon; the most frequent and brightest they are in the auroral oval, the belt surrounding the polar region.

When thinking of the arctic and the north in a more practical manner, the individual boundaries are not crucially important. The matter of primary importance is the basic character of those areas; the core, where several of the criteria presented exist.

In addition to scientific definitions there are political definitions for the north and the arctic. The European Union's Northern Dimension is such a type of political definition. It is a geopolitical definition related to the point of the compass and refers to the European Union's northern periphery. Its specific aim is to raise the European Union's profile in Northern Europe. The concept of the Arctic Environment occurs in many contexts, such as in names of institutes and journals.

The arctic and the high northern areas are unique and vulnerable. Living there and sustainable use of the areas' natural resources, such as oil, gas, industrial metals, game and fish demand knowledge and a continuous scientific research of physical, biological and social conditions.

The high latitudes were first inhabited by indigenous peoples with their own livelihoods. For instance, in northern Fennoscandia, reindeer husbandry has been traditionally practised by the Sámi people. As newcomers, people of a European background have changed local traditions. Those researching human activities in the north face these days the problems and conflicts with native people struggling to have the right for their traditional use of the land and waters. Intruding cultures have not only brought welfare, but also destructive habits and diseases. Our highly technocratic civilization has exploited natural resources and is responsible for the many indirect negative impacts on the remote northern regions and the people living there. For example, air-borne pollutants and toxic metals spoil the arctic environment even if industrial activities take place far from these areas. This is due to air currents transporting emissions over long distances from the areas of production to the far north. It is also true that the arctic itself is increasingly industrialized. Also the growing interest of tourism to the northern and arctic regions has become a concern.

Increasing greenhouse effects and global warming with the associated environmental changes is of great concern. Perhaps of equal concern are also the new opportunities for the ecosystems and life, particularly at the high latitudes.

At the University of Oulu there is a unit called the Thule Institute. It focuses on Northern and Environmental Issues. The Thule Institute operates as a store of national and international expertise in the following research programmes: Global Change in the North, Northern Land Use and Land Cover, and Circumpolar Health.

New Applications of Political Ecology for the Nordic Context

Paul Robbins[1] & Hannu Heikkinen[2]
[1] University of Arizona, Department of Geography and Regional Development
[2] University of Oulu, Department of Art Studies and Anthropology

Abstract: Recent developments in the field of political ecology have resonance for themes in northern Geography. In this review, we briefly summarize areas of emerging theory in the field, pointing to the way Political Ecology has come to explore difficult questions for reconciling constructivist accounts of environmental change with material ecologies and for understanding and thinking about scientific expertise and technologies of measurement. Briefly connecting these theoretical problems with cases of environmental change, conservation, and forestry from the Finnish context, we suggest advances in political ecology open new questions and new empirical research problems for the Nordic context.

The politics of the environment is by no means a new problem, nor one unique to recent academic inquiry. Struggles over resources and the control of nature are classic themes in the fields of anthropology, history, and especially geography, where human/environment interactions take centre stage in analysis. Political ecology as a mode for explaining and understanding environmental conditions and change is a well-established approach. As "empirical, research-based explorations to explain linkages in the condition and change of social/environmental systems, with explicit consideration of relations of power" (Robbins 2004: 12), political ecology has been used to explain land degradation, explore regional environmental struggles, and reveal the root failings of conservation policy. The application of this approach to the Nordic context has revealed the way, for example, energy development in Norway enflames struggles between Sámi (Lapp) herders and hydropower (Dalland 2006), and the way political economic transition of the forestry sector in Finland has led to new management regimes and potentially new forest ecologies (Lehtinen 2001).

New problems in political ecology

Recent years have brought new problems and questions to the field of political ecology, however, presenting researchers both with new tools but also with new conceptual puzzles. Specifically we suggest, recent advances in political ecology have forced difficult choices for reconciling constructivist accounts of environmental change with material ecologies and for understanding and thinking about scientific expertise and technologies of measurement.

Reconciling constructivism and the material world

Political ecological research that critically interrogates environmental change has frequently led to the revelation that taken-for-granted accounts of environmental conditions or processes are often unreliable and are heavily influenced by the political and economic histories of the places in which they are imposed. Fundamental categories, including concepts like "wilderness" and "pristinity" have been shown to be historically contingent and culturally dependent, rooted in colonial, nationalist, and market-based discourses (Neumann 1998). In West Africa for example, long held assumptions of deforestation has recently been questioned through interrogation of flimsy baseline data used to estimate change. The willingness of colonial officers and contemporary managers to believe unstable assumptions, however, was heavily influenced by the authority these *discourses* have in the larger historical political economy of forest control (Fairhead & Leach 1996).

This approach to environmental politics has been useful, but has led to a recent round of debate and reconsideration. Critically evaluating, deconstructing, and dismantling the environment as a discursive construction raises questions about the ability to make grounded claims about material conditions as they "actually" are. It becomes more difficult to defend claims about environmental conditions and change while simultaneously undermining the tools and viewpoints that allow an objective view of those very conditions.

The reconciliation of this problem, or at least the path towards taking both claims seriously – that environments change but that claims of environmental change are social constructions – have occupied a recent generation of political ecologists. Offering research protocols that take seriously the influence of non-human actors (trees, soils, or climate) on socio-political systems, political ecology has begun interrogating the line between discourse and the material world (Bakker & Bridge 2006).

Querying science and technology

In a related way, political ecological understanding of science and technology has been influenced by innovations from other fields, especially the Sociology of Scientific Knowledge (SSK). For researchers in this field, scientific claims have histories, social conditions under which they are produced, and economic contexts that condition the way research about them is performed. While by no means espousing a totally relativistic position on the nature of "truth", such an approach admits that science is a social effort, undertaken by living people with their own gendered, classed and raced status, within institutions that must be funded and that have their own habits (Merton 1973; Latour 1987; Haraway 1989).

Technologies for monitoring and managing the environment have been treated with the same critical scrutiny. To what degree can we understand technology to have its own momentum, to not only reflect the needs or desires of a society, but to actually change these needs, as people and communities *reverse adapt* themselves to fit the metrics of their tools (Winner 1977)? To the benefit and loss of whom? And with what ecological impact?

Consider the case of remote sensing of forest cover change in India. While the tool is crucial to assessing environmental conditions, it tends to lead to management decisions that change the character of ecosystems. Specifically, sensitivity on the part of forest managers to the way the tool measures forest cover and its horizontal spread, leads them to plant, encourage, or tolerate fast-growing species that are easily visible from the air (Robbins 2001). The technology, however important, is embedded in the social and scientific systems it is designed to measure, and is never innocently objective in this sense.

New political ecologies of nordic environments

Applying these perspectives in political ecology to the Nordic setting, new empirical research questions come into view. Consider Sámi reindeer herding in Northern Finland and its interaction with conservation and forestry efforts. The imposition of a strict nature reserve at Malla in the municipality of Enontekiö (an IUCN category Ia reserve to protect biodiversity) has entailed the exclusion of human influences, specifically herding in the reserve. Local Sámi access rights were transferred to other groups, specifically scientists and tourists, without any form of compensation (Jokinen 2005). This is a familiar political ecology story for many indigenous societies throughout the world.

A further problem is suggested in this effort, however, that being the divergent definitions of the "natural" system state, and the relationship between the constructed character of the Malla reserve and its material condition. Interdisciplinary research at the reserve has shown that *both* grazing and the prohibition of grazing create selective pressures that favour some species over others, leading to divergent ecologies. Strict protection selects against some small, Calcium-favouring, short lifespan seed plants, which are often red listed in Finland, like *Viola rupestris* ssp. *relicta*, (and related beetle, butterfly, spider etc.). These communities, have evolved under grazing and trampling. Strict protection conversely favours some lush perennial vegetation like *Angelica archangelica* or *Cicerbita alpina* (L.) Wallr. (and related other species), which on the other hand, meet certain aesthetic expectations and imaginaries, especially those tied to a "humanless - edenic" natural state (Jokinen 2005; Olofsson & Oksanen 2005). This issue is fundamentally discursive, since landscape expectations and imaginaries are human *constructions*. The way these sculpt new *material* forest ecologies, often with unintended consequences, remains a less well examined phenomenon.

So too, while the science used in monitoring biodiversity in the reserve is rigorous, technically elegant, and intrinsically objective, it is never innocent of the metrics and decisions imposed prior to assessment. While for herders biodiversity means certain holistically perceptible and historically evolved states of their pastures, for specialists it means certain states and relationships comprised of specific species of interest (Jokinen 2005). These divergent understandings also influence the choices of baseline data used to understand change, the acceptance of certain forms of observational data as reliable or unreliable, and the definition of system states as "diverse". This is made more problematic by the numerous technical definitions and measures of biodiversity itself, and the problems associated with obtaining robust measures (Dobson 2005). In this sense, much more needs to be known concerning the way scientific monitoring of the reserve is actually carried out by experts, the technologies of monitoring they use, and the underlying assumptions and definitions they bring to bear on the problem. In other words, to understand the problem at Malla, conservation biologists in Finland require the kind of ethnographic attention previously reserved for Sámi herders.

Similarly, Sámi herders have long struggled with the state Metsähallitus company in the Municipality of Inari in Finland, claiming that forest development is destroying last vital winter pastures of reindeer (Kyllönen & Raitio 2004; Hyppönen 2005). Such decisions have been defended by claims of the relatively more valuable contribution of forestry and tourism to the regional economy. The measures and metrics underlying these claims, however, demand more scrutiny than they have to date received. Sampling of tourism data follows regional districts and borders of municipalities that match neither the areas of reindeer management nor forestry. Economic data are also temporally mismatched, following an official year calenderer starting the 1st of January while "the reindeer herding year" begins 1st of June when most new calves are born. The result is an unclear comparison of relative contributions and flows of value in particular regions and during particular seasons (Heikkinen 2002; Kyllönen & Raitio 2004; Hyppönen 2005, 2006; Raitio 2006). Currently it is openly admitted that herding is important for tourism and vice versa, but the struggle over what that importance is and how it will be fostered or maintained is afloat in a sea of numbers. The techniques of environmental and economic assessment, therefore, demand the attention of political ecologists in the Northern context as much as the practices and politics of indigenous or local peoples.

In sum, environmental facts and statistical data, while essential for management, are created under conditions that are inevitably social and political, within expert communities with their own ideas and discourses. These statistics and measures, moreover, must be interpreted and implemented within further fields of political struggle. While "knowledge is power", it is also conversely clear that power produces knowledge and sets the conditions for what claims are understood to be true (Foucault 1980). Discourses, expert communities, and technologies of measurement are as much a part of political ecology as systems of production, land rights, and resource use. Their incorporation into Northern Geographies in the future will propel new and important research questions.

References

Bakker, K. & G. Bridge (2006). Material worlds? Resource geographies and the 'matter of nature'. *Progress In Human Geography* vol. 30:1, 5-27.

Dalland, O. (2006). Science, local knowledge and exclusionary practices: Lessons from the Alta Dam case. *Norsk Geografisk Tidsskrift* vol. 60:2, 149-160.

Dobson, A. (2005). Monitoring global rates of biodiversity change: challenges that arise in meeting the Convention on Biological Diversity (CBD) 2010 goals. *Philosophical Transactions Of The Royal Society B: Biological Sciences* vol. 360:1454, 229-241.

Fairhead, J. & M. Leach (1996). *Misreading the African Landscape: Society and Ecology in a Forest-Savanna Mosaic.* Cambridge University Press, Cambridge.

Foucault, M. (1980). Truth and Power. *In* C. Gordon (ed.): *Power/Knowledge: Selected Interviews and Other Writings 1972-1977*, 109-133. Pantheon, New York.

Haraway, D. (1989). *Primate Visions: Gender, Race, and Nature in the World of Modern Science.* Routledge, New York.

Heikkinen, H. (2002). Sopeutumisen mallit: Poronhoidon adaptaatio jälkiteolliseen toimintaympäristöön Suomen läntisellä poronhoitoalueella 1980-2000. Art Studies & Anthropology, University of Oulu (SKS).

Hyppönen, M. (2005). Metsänkäyttömuotojen yhteensovittamisen ongelmat Ylä-Lapissa. *In* E. Lindqvist & P. Posio (eds.): *Lapin Natura 2000 -artikkelisarja*, 32-33. Lapin ympäristökeskuksen moniste 61.

Hyppönen, M. (2006). Inarin metsäkiista tutkimuksen kohteena. *Lapin Tutkimusseuran vuosikirja* 44-45, 6-13.

Jokinen, M. (ed.) (2005). Poronhoidon ja suojelun vaikutukset Mallan luonnonpuistossa. Metsäntutkimuslaitoksen tiedonantoja, Finnish Forest Research Institute (METLA).

Kyllönen, S. & K. Raitio (2004). Ympäristöristiriidat ja niiden hallinta: puuntuotannon ja poronhoidon välinen konflikti Inarissa. *Alue ja Ympäristö* vol. 33: 2, 3–20.

Latour, B. (1987). *Science in Action: How to Follow Scientists and Engineers through Society.* Harvard University Press, Cambridge.

Lehtinen, A. A. (2001). Globalisation and the Finnish forest sector: On the internationalisation of forest industrial operations. *Fennia* vol. 179: 2, 1-2.

Merton, R. K. (1973). *The Sociology of Science.* University of Chicago Press, Chicago.

Neumann, R. P. (1998). *Imposing Wilderness: Struggles over Livelihood and Nature Preservation in Africa.* University of California Press, Berkeley.

Olofsson, J. & L. Oksanen (2005). Effects of reindeer density on vascular plant diversity on North Scandinavian mountains. *Rangifer* vol. 25: 1, 5-18.

Raitio K. (2006). Environmental conflicts and their management: the conflict between reindeer herding and timber production in Inari, Finland. *Presented paper at the conference "Förvaltning av omstridda naturresursser. Nordiska trender och nya utmaningar" at the University of Umeå, Sweden, March 15-16, 2006.*

Robbins, P. (2001). Fixed Categories in a Portable Landscape: the Causes and Consequences of Land Cover Classification. *Environment and Planning A* vol. 33: 1, 161-179.

Robbins, P. (2004). *Political Ecology: A Critical Introduction.* Blackwell, New York.

Winner, L. (1977). *Autonomous Technology: Technics-out-of-control as a Theme in Political Thought.* MIT Press, Cambridge.

The Potentiality of Remote Sensing in Biogeographical Research

Miska Luoto
Thule Institute, University of Oulu

Abstract: Spatial variability in species' geographical patterns is a crucial phenomenon and has formed the basis for numerous biogeographical studies. At present, human activities increasingly affect the terrestrial biosphere, resulting in habitat loss and degradation. Due to the rapidity of this process there is utmost need to detect and predict changes in the natural environment and assess the spatial distribution of valuable sites and habitats. Remote sensing (RS) produces valuable information for species mapping and monitoring. Due to their high spatial and temporal coverage, RS data have clear advantages compared to any other source of information. However, the benefits of remote sensing have not been fully utilized in biogeographical studies, and several methodological issues remain insufficiently explored. This paper presents novel approaches and techniques to integrate RS and geographic information (GI) data into biogeographical research. The issues discussed in this paper can have relevance in several fields of application of geographical data.

Introduction

Which factors explain the distribution and the abundance of a species has continued to be a central question in biogeographical research since the pioneering work of Andrewartha & Birch (1954). Nowadays, human activities increasingly affect the terrestrial biosphere, resulting in habitat loss and degradation which ultimately impair ecosystem function and ecosystem services (Matson & al. 1997). Due to the rapidity of this process, there is utmost need to detect and predict changes in the natural environment and assess the spatial distribution of valuable sites and habitats.

Current European and national environmental regulations often require assessments of the ecological effects of land use planning (Luoto & al. 2002b). Not only detailed knowledge of the present patterns of species richness and distributions (hereafter biodiversity), but also accurate spatial predictions for more poorly known regions are needed. The challenge in conservation and land use planning is that biodiversity data, as accurate as possible, is required within a limited amount of time and over considerably large areas. Unfortunately, such data are often sparse and expensive to acquire by traditional field inventories.

Alongside with biogeographers and landscape ecologists, environmental researchers, planners and policy-makers are increasingly turning to the discipline of remote sensing (RS) and geographic information (GI) science. RS-GI science is considered to be capable of providing the required techniques and data for developing cost-effective assessments, models and projections of biodiversity responses to environmental change over large areas. The expectations on RS-GI techniques have been boosted by the growing selection of remotely sensed data (often freely accessible) sources. At the moment, data archives provide increasing quantities of multi-temporal and multi-spectral remotely sensed imagery (Luoto & al. 2002b).

Remote sensing generates a remarkable array of biogeographically valuable measurements as well as the capacity to detect natural and human-induced land cover changes. Used in an integrative mode, RS-GI data can provide information about both historical and current land cover factors affecting biodiversity patterns. Relationships between species distributions and remotely sensed data, particularly in conjunction with topographic, soil and bedrock information, can be used to predict the distribution of species over large areas.

Remotely sensed imagery exposes land cover changes at spatial scales from local to continental, allowing one to monitor the pace of habitat conversion. These measurements of habitat loss can be used in providing quantitative estimates of biodiversity loss through the use of species-habitat modelling. The challenge is to go one step further and provide a more detailed understanding of which species are being lost and why. How can we best match existing and emerging RS and GI technologies to parameters that have clear implications for organisms and ecosystems?

Hitherto, the benefits of remote sensing have not been fully utilized in biodiversity studies. Moreover, there are several critical issues in applying RS-GI based biodiversity models which have been insufficiently explored. These include the applicability of different RS data sets in biodiversity assessments, and the effect of scale, statistical techniques and model complexity on RS-GI based biodiversity modelling. In this article, I present novel approaches and techniques to integrate RS and geographic information (GI) data into biogeographical research. I wish that the issues discussed in this paper can have relevance in several fields of application of spatial data in physical geography and related environmental sciences.

The potential of remote sensing for assessing biodiversity

Current approaches in using RS in biodiversity modelling can be categorized into two main types: 1) habitat mapping using RS data, and predictions of species distribution based on habitat requirements; and 2) direct relationships between spectral radiance values recorded from remote sensors and species' distributions recorded in field surveys. These approaches require the use of information collected from the field, as well as from the remote sensor. It is essential to establish the strength of the linkages between these different kinds of information collected on different scales. For example, the degree of correspondence between habitat maps and species distributions depends on the degree of habitat map generalization. Such relationships should be optimized to get maximum information on biodiversity. In comparison, a major limitation of spectral values-biodiversity models is that the relationship between spectral reflectance and species distributions depends not only on the species in question, but also the area. In practice, little guidance exists for investigators concerning the suitability of these methods for biodiversity assessments in different habitats. Thus, there is an utmost need to evaluate the applicability of these two main types of RS information in modelling biodiversity on different spatial scales in different landscape mosaics. I propose that such comparative work is greatly needed to produce guidelines for choosing optimal approaches in RS-GIS based biogeographical modelling.

The effect of scale on the biogeographical models

The problems and impacts of scale have long been a central issue in biogeographical (Rahbek & Graves 2001; Blackburn & Gaston 2002), geomorphological (Walsh & al. 1998; Luoto & Hjort 2006) and landscape ecological studies (Wiens 1989; Levin 1992). The spatial scale on which species distribution modelling is undertaken is of fundamental importance for the results and inferences of biogeographical studies. The concept of spatial scale consists of two important attributes: the unit of sampling and the geographical space covered. The first attribute is defined by 'grain' (or 'resolution') and 'focus', grain being the size of the common analytical unit and focus the area represented by each data point. The second attribute is 'extent', describing the geographical area over which comparisons are made (Scheiner 2003; Rahbek & Graves 2001).

Most importantly, the choice of spatial resolution in modelling exercises can directly affect the ecological interpretations derived from the results and the usability of RS-GI based models. If the response of a given species to environmental factors shows a high scale-sensitivity, comparative studies over a range of scales can indicate the resolutions at which the species-environment patterns are most accurately modelled. However, we lack the knowledge of how the extent of habitat information and resolution of input data affect the performance of RS-GI based habitat-relationship models on different spatial scales.

Researchers have come to different conclusions about the usefulness of habitat-relationship models for predicting species presence or absence. This difference very likely stems from the failure to recognize the effects of spatial scale. In order to increase the plausibility of the species distribution models, the analysis should be conducted on the scale on which the target species most strongly responds to the environmental variation (Wiens 1989; Carroll & al. 1999). Identification of the most suitable resolution for species distribution modelling often requires studies conducted over a range of spatial scales. Krawchuk & Taylor (2003) argued that studies examining species response to the environment performed at a single scale may result in limited conclusions and potentially harmful management recommendations.

The effect of modelling techniques and complexity on RS-GI based models

Recently, species distribution modelling has become one of the key issues in biogeography

(Luoto & al. 2002a; Thuiller 2003), conservation planning (Ferrier & al. 2002; Lehmann & al. 2002) as well as in landscape ecology (Fleishman & al. 2001; Luoto & al. 2002b; Lawler & al. 2004). This development is based on two trends: growth in the availability of remotely sensed (RS) data and development of GI techniques (Gould 2000; Nagendra 2001; Kerr & Ostrovsky 2003), and the development of novel statistical techniques that can be applied to a wide range of ecological systems (Guisan & Zimmermann 2000; Thuiller 2003).

RS-GI based biodiversity modelling studies have usually been conducted by employing only one modelling approach. Obvious variation is expected from using different techniques because different models use a variety of assumptions, algorithms and parameterizations. Thus, when studies use a single modelling technique there is no information of whether the selected method provides the best predictive accuracy for the particular data set used. Only a few attempts have been made to evaluate the ability of different methods to identify plausible and accurate species-environment responses in complex RS-GI multiple regression settings including many potentially important explanatory variables (Guisan & Zimmermann 2000; Thuiller 2003). Consequently, insufficient guidance exists for investigators concerning the suitability of different modelling approaches for species distribution modelling and other biodiversity assessments.

Currently, there are several different modelling approaches available for biogeographical modelling, ranging from generalized linear models (GLM) and generalized additive models (GAM) (Bustamante 1997; Seoane & al. 2003; Luoto & al. 2004), to rule-based techniques (artificial neural networks; ANN and classification tree analysis; CTA) (Moisen & Frescino 2002; Olden & al. 2004), and novel regression techniques, for example multivariate adaptive regression splines (MARS) (Venables & Ripley 2002). Generalized linear models are mathematical extensions of linear models which can handle non-linear relationships and different types of statistical error distributions such as Gaussian, Poisson, Binomial or Gamma (Crawley 1993). Generalized additive models are flexible data-driven nonparametric extensions of generalized linear models (Hastie & Tibshirani 1990) that allow both linear and complex additive response curves to be fitted. Classification tree analysis is a rule-based method which has rarely been applied in RS-GI based modelling studies. CTA does not rely on a priori hypotheses about the relationships between predictor and response variables. The method generates a binary tree through binary recursive partitioning, a process that splits a node based on true/false answers about the values of predictors (Venables & Ripley 2002). The rule generated at each step maximizes the class purity within each of the two resulting subsets. Artificial neural networks are complex non-parametric techniques that have been shown to provide highly flexible function approximates for any data (Lek & Guegan 1999). Neural networks can provide a means to build accurate models when the functional form of the underlying equations is unknown (Ripley 1996). Multiple adaptive regression splines is a relatively new technique that combines classical linear regression, mathematical construction of splines and binary recursive partitioning to produce a local model in which relationships between responses and predictors are either linear or non-linear

(Friedman 1991). MARS has been advocated as an effective method to model complex spatial data that can include a relatively high number of variables.

My earlier results (Luoto & Hjort 2005) as well as the studies by Moisen and Frescino (2002), Thuiller (2003) and Segurado and Araújo (2004) suggest that the use of novel non-parametric techniques often results in a slightly better model performance than the parametric approaches. Non-parametric modelling techniques are powerful tools with limited statistical theory to support them, and investigators should be aware of their potential shortcomings. Non-parametric techniques have been criticized because they can produce a complex set of functions, whereas a corresponding parametric function may capture most of the same variation (Venables & Ripley 2002).

Downscaling species atlas distributions

One of the limitations in using biogeographical data sets, typically species' distribution atlases, in regional planning is their coarse resolution (typically from 10-km to 50-km) relative to the requirements of conservation and land use planning. In response to this, attempts have been made to downscale species distributions to finer resolutions (McPherson & al. 2006). The term 'downscaling' describes a procedure in which information about a process with a certain characteristic scale is derived from other processes with larger scales. The general limitations, theory and practise of downscaling methods are well described in the literature (McPherson & al. 2006). However, only little empirical work has been conducted in geography to use coarse-grained data in order to model biogeographical processes at finer spatial resolution, however see exceptions in Araújo & al. (2005) for species distribution modelling in U.K. Systematic studies employing downscaling methods for several different biogeographical processes utilizing independent model calibration and model test areas are largely lacking. By and large, the potentiality of statistical downscaling of the distribution of flora and fauna has been insufficiently investigated.

Downscaling has been done by combining remotely sensed data and expert opinion to assign species to habitats or land-cover classes considered suitable for species. One of the problems with this approach is that there may be insufficient expert knowledge of species-habitat relationships for many species and areas. An alternative is to use empirical modelling techniques that explore the correlation between species distribution and sets of predictor variables to downscale distributions of species to finer spatial resolution. One crucial assumption in this approach is that the main drivers of species distribution at coarse resolution are also important determinants of distributions at finer resolutions.

Recent studies have demonstrated significant variability in downscaled model predictions (Araújo & al. 2005; McPherson & al. 2006). These results highlight the need for further research to test and improve the approach so as to increase confidence in model predictions. Statistical downscaling has important limitations, and departures from model assumptions are likely to vary for different species and regions. In particular, it would be important to investigate which species are poorly modelled, where and why. Without such an assessment it is difficult thoroughly to assess the usability and accuracy of downscaled models in real-world situations.

Conclusions

Remote sensing and GI techniques provide an approximate cost-efficient method to estimate the biodiversity status of wide areas on a broad scale. I put forward that the RS-GIS based modelling approaches deserve further attention in biogeography and applied land use planning. However, there are also many potential shortcomings in using satellite imagery and other kinds of spatial data as a surrogate for species richness – how to appropriately collect source data for modelling, technical problems in processing satellite imagery, methodological pitfalls in regression models, etc. The integration of biogeographical modelling, RS and GIS technology requires often transdisciplinary skills between geography, ecology, statistics and geoinformatics. Thus the pitfalls for the misuse and abuse of the GIS technology with high calculation capacity are very obvious. We should be quite cautious and not apply the results of predictive models and extrapolations too uncritically, and not without a good geographical knowledge of the study area. In sum, broad-based understanding of geoinformatics and related issues is needed by spatial data producers and modellers.

References

Andrewartha, H.G. & L.C. Birch (1954). *The distribution and abundance of animals.* The University of Chicago Press.

Araújo, M.B., W. Thuiller, P.H. Williams & I. Reginster (2005). Downscaling European species atlas distributions to a finer resolution: implications for conservation planning. *Global Ecology and Biogeography*, vol.14, 17-30.

Blackburn, T. & K. Gaston (2002). Scale in macroecology. *Global Ecology & Biogeography*, vol. 11, 185-189.

Bustamante, J. (1997). Predictive models for Lesser Kestrel Falco naumanni distribution, abundance and extinction in southern Spain. *Biological Conservation*, vol. 80, 153-160.

Carroll, C., W.J. Zielinski & R.F. Noss (1999). Using presence-absence data to build and test spatial habitat models for the fisher in the Klamath Region, U.S.A. *Conservation Biology*, vol. 13, 1344-1359.

Crawley, M.J. (1993). *GLIM for ecologists.* Blackwell Scientific Publications, Oxford.

Ferrier, S., G. Watson, J. Pearce & M. Drielsma (2002). Extended statistical approaches to modeling spatial pattern in biodiversity: the north-east New South Wales experience. I. Species-level modeling. *Biodiversity and Conservation*, vol. 11, 2275-2307.

Fleishman, E., R. Mac Nally, J.P. Fay & D.D. Murphy (2001). Modelling and predicting species occurrence using broad-scale environmental variables: an example with butterflies of the Great Basin. *Conservation Biology*, vol. 15, 1674-1685.

Friedman, J. (1991). Multivariate adaptive regression splines. Annals of Statistics, vol. 19, 1-141.

Gould, W.A. (2000). Remote sensing of vegetation, plant species richness, and regional diversity hotspots. *Ecological Applications*, vol. 10, 1861-1870.

Guisan, A. & N.E. Zimmermann (2000). Predictive habitat distribution models in ecology. *Ecological Modelling*, vol. 135, 147-186.

Hastie, T. & R. Tibshirani (1990). *Generalized additive models.* Chapman and Hall, London.

Kerr, J. & M. Ostrovsky (2003). From space to species: ecological applications for remote sensing. *Trends in Ecology and Evolution*, vol. 18, 299-305.

Krawchuk, M. & P. Taylor (2003). Changing importance of habitat structure across multiple spatial scales for three species of insects. *Oikos*, vol. 103:1, 153-161.
Lawler, J., R. O'Connor, C. Hunsaker, K. Jones, T. Loveland & D. White (2004). The effects of habitat resolution on models of avian diversity and distributions: a comparison of two land-cover classifications. *Landscape Ecology*, vol. 19, 515-530.
Lehmann, A., J. Overton & J. Leathwick (2002). GRASP: generalized regression analysis and spatial prediction. *Ecological Modelling*, vol. 157, 189-207.
Lek, S. & J. Guegan (1999). Artificial neural networks as a tool in ecological modelling, an introduction. *Ecological Modelling*, vol. 120, 65-73.
Levin, S.A. (1992). The problem of pattern and scale in ecology. *Ecology*, vol. 73, 1943-1967.
Luoto, M. & J. Hjort (2005). Evaluation of current statistical approaches for predictive geomorphological mapping. *Geomorphology*, vol. 67, 299-315.
Luoto, M. & J. Hjort (2006). Scale matters – a multi-resolution study of the determinants of patterned ground activity in subarctic Finland. *Geomorphology*, vol. 80, 282-294.
Luoto, M., M. Kuussaari & T. Toivonen (2002a). Modelling butterfly distribution based on remote sensing data. *Journal of Biogeography*, vol. 29, 1027-1037.
Luoto, M., T. Toivonen & R.K. Heikkinen (2002b). Prediction of total and rare plant species richness in agricultural landscapes from satellite images and topographic data. *Landscape Ecology*, vol. 17: 3, 195-217.
Luoto, M., R. Virkkala, R.K. Heikkinen & K. Rainio (2004). Predicting bird species richness using remote sensing in boreal agricultural-forest mosaic. *Ecological Applications*, vol. 14, 1946-1962.
Matson, P.A., W.J. Parton, A.G. Power & M.J. Swift (1997). Agricultural intensification and ecosystem properties. *Science*, vol. 277, 504-509.
McPherson, J., W. Jetz & D. Rogers (2006). Using coarse-grained occurrence data to predict species distributions at finer spatial resolutions - possibilities and limitations. *Ecological Modelling*, vol. 192, 499-522.
Moisen, G. & T. Frescino (2002). Comparing five modelling techniques for predicting forest characteristics. *Ecological Modelling*, vol. 157, 209-225.
Nagendra, H. (2001). Using remote sensing to assess biodiversity. *International Journal of Remote Sensing*, vol. 22, 2377-2400.
Olden, J.D., M.K Joy & R.G. Death (2004). An accurate comparison of methods for quantifying variable importance in artificial neural networks using simulated data. *Ecological Modelling*, vol. 178, 389-397.
Rahbek, C. & G.R. Graves (2001). Multiscale assessment of patterns of avian species richness. *Proceedings of the National Academy of Sciences, USA*, vol. 98, 4534-4539.
Ripley, B. (1996). *Pattern recognition and neural networks*. Cambridge University Press, New York.
Scheiner, S.M. (2003). Six types of species-area curves. *Global Ecology and Biogeography*, vol. 12, 441-447.
Segurado, P. & M. Araújo (2004). An evaluation of methods for modelling species distributions. *Journal of Biogeography*, vol. 31, 1555-1569.
Seoane, J., J. Vinuela, R. Díaz-Delgado & J. Bustamante (2003). The effects of land use and climate on red kite distribution in the Iberian peninsula. *Biological Conservation*, vol. 11, 401-414.
Thuiller, W. (2003). BIOMOD - optimizing predictions of species distributions and projecting potential future shifts under global change. *Global Change Biology*, vol. 9, 1353-1362.
Walsh, S.J., D.R. Butler & G.P. Malanson (1998). An overview of scale, pattern, process relationships in geomorphology: a remote sensing and GIS perspective. *Geomorphology*, vol. 21, 183-205.
Venables, W.N. & B.D. Ripley (2002). *Modern applied statistics with S*. Springer-Verlag, Berlin.
Wiens, J.A. (1989). Spatial scaling in ecology. *Functional Ecology*, vol. 3, 385-397.

Combining Data from Satellite Images and Reindeer Herders in Arctic Petroleum Development: the Case of Yamal, West Siberia

Timo Kumpula[1], Bruce Forbes[2] & Florian Stammler[2,3]
[1] Department of Geography, University of Joensuu
[2] Arctic Centre, University of Lapland
[3] Scott Polar Research Institute, University of Cambridge

Abstract: The aim of the study is to assess the capacity for satellite imagery in detecting different natural and anthropogenic land cover types in the vicinity of a modern petroleum extraction development in the Russian Arctic. The Yamal Peninsula in northwest Siberia contains some of the largest untapped deposits known in the world. It also serves as the homeland of the Yamal Nenets, who have exploited first wild and then domestic reindeer in the region for at least 1000 years. Their annual migration from the treeline to the northern tundra brings them into contact with a number of impacts associated with gas exploration and production. These range widely and include physical obstructions from roads, railways, and pipelines, as well as direct and indirect ecological impacts, such as changes in vegetation, soils and hydrology due to e.g. drilling, infrastructure development, and seismic surveys. Some of the effects are relatively small-scale, only a few meters across, while others cover several hectares. Nenets' perceptions of the spatial aspects of their territories encompass changes in both quantity and quality of terrestrial habitats, rivers, lakes and campsites that have been used seasonally for centuries. Even high-resolution imagery was unable to detect things like trash (rusted metal, broken glass), drilling muds and petro-chemicals that can strongly affect the overall quality of reindeer pastures. To properly assess the overall ecological impacts of petroleum development requires a combination of state-of-the-art remote sensing coupled with detailed ground-truthing. These efforts must embrace both scientific and local knowledge from indigenous herders and also non-indigenous gas field workers.
Key words: Nenets, reindeer migration, indigenous knowing, remote sensing, arctic tundra

Introduction

Russia is becoming more important as a producer and supplier of oil and gas to European and eventually North American markets. The Iraq war and post-war chaos have partly increased world market price of oil barrel to its record. Meanwhile Russia has invested in pipelines across eastern and central Europe and tanker traffic via the Northern Sea Route. Among the main sources are the giant oil fields of the Nenets Autonomous Okrug (NAO), Khanti-Mansisk Autonomous Okrug (KMAO), and the super giant gas fields of the Yamal-Nenets

Autonomous Okrug (YNAO). Russian production has an image of being more stable and safe from political crisis than some of the big middle Eastern producers. It has once again become a powerful trading partner, in particular as a major source of energy for the global market. At present Russia supplies 25% of the world's natural gas, which makes it the No. 1 producer. Some 90% of this production comes from West Siberia. As for oil, Russia recently also took over the leading position in world production from Saudi Arabia (Mosnews 2006). 60% of Russia's oil is from West Siberia. Overall EU dependency on both oil and gas from Russia is high and increasing, relative to European sources like Norway and the declining export capacity of the UK. In 2005 the European Union received 25% of its oil (30% of all imports) and 25% of its gas (50% of all imports) from Russia. Dependency on Russia among individual EU countries, however, varies greatly. For example, Finland relies 100% on Russian gas sources, whereas France with its traditional ties to Algeria imported only 24% from Russia as of 2004.

As petroleum industrial infrastructure expands rapidly throughout Russia's north, it has profound implications for the environment and region's economies and local people, in particular indigenous groups practicing traditional livelihoods like herding domestic reindeer (*Rangifer tarandus*), hunting and fishing. The building up of the Russian empire has affected these indigenous groups over the last 500 years. Up until the 19th century the main resource produced from these northern regions was fur, whereas from the mid-20th century onward petroleum products became the central focus.

The exploration and exploitation phases of northern areas like YNAO and NAO were launched in the 1960's with extensive resource prospecting. Geological surveys searched through the tundra zone with heavy drilling equipment and left clear imprints on the permafrost landscapes. Surveys mapped the oil and gas deposits and drilling was first started in boreal and subarctic areas with easiest access from the south, such as existing railways and roads. Surveys and drilling further north in truly arctic areas required enormous amounts of infrastructure for example building of pipelines, expanded road and rail networks and accommodation facilities. Supply of remote arctic exploration sites has been difficult with winter road networks, helicopter, aeroplane and also by ships via the Northern Sea route. In the post-Soviet period, oil and gas have become a vital source of profits to the overall Russian economy. The exploration and exploitation phases of petroleum development result in benefits and wealth but also, at the same time, in a large number of direct and cumulative impacts on ecosystems and cultures all along the route from the source to the market (Forbes 2004; Stammler & Wilson 2006). New drilling sites and oil and gas fields have been connected to transport networks to export oil and gas to world markets. This of course has increased impacts on the environment and local indigenous communities.

In the YNAO indigenous lifeways of the Nenets, Khanty, Komi-izhemtsy and Sel'kup fishermen, reindeer herders and hunters - for the many of whom are still nomadic or semi-nomadic – are comparatively well-preserved. While use of reindeer reaches back to at least 1200 years, the large scale herding of domestic reindeer has developed in the 1600's. In YNAO on Yamal Peninsula reindeer herding survived intact from the Soviet period with the least amount of damage compared to other indigenous cultures elsewhere in Russia (Stammler 2005).

Satellite images have been used successfully in the detection of marine and coastal oil spills. Especially radar satellite images are suitable for detection of oil spills (Benelli & Garzelli 1999; Espedal & Johannessen 2000; Jones 2001; Solberg & al. 2003). Oil spill detection from terrestrial drilling sites is more difficult, because the size of spill is an important factor that defines its potential for detection. In detection of impacts not only spills should be detected. Impacts should be studied holistically, all possible impacts on environment and other land use forms. Also impacts on the herding community must be studied, and several authors have argued for a shift in the focus of impact studies from spill response and impact and damage compensation to prevention of spills, minimising negative impacts and development plans for local populations (Meschtyb & al. 2005; Stammler & Wilson 2006).

Impacts of industrialisation in the northern Russian mining towns of Vorkuta and Noril'sk have been studied with Landsat TM data (Toutoubalina & Rees 1999; Virtanen & al. 2002). Tømmervik & al. (2003) studied effects of air pollution on vegetation in the Norwegian-Russian border area by using multitemporal Landsat MSS/TM data. Stow & al. (2004) reviewed the possibilities of change detection studies in arctic areas with available remote sensing data. Rees & al. (2003) studied land cover change in NAO with Landsat TM and ETM+ images, they also attempted to incorporate indigenous knowledge of Nenets reindeer herders to land cover change interpretation, although this formed only a very small part of their analysis. One problem associated with such attempts is that indigenous knowledge is highly situative, very detailed, learned and enacted by people moving through the land. Prominent scholars have therefore questioned the compatibility of "indigenous knowledge" with natural science knowledge that can be put into databases and treated as body of information independent from its context, whereas indigenous ways of knowing are culturally embedded sets of practices (Ingold 2005; Stammler 2005). Landsat MSS/TM/ETM+ data are too coarse to detect small-scale impacts that are common in most cases. Very high resolution images like IKONOS-2 and Quickbird-2 have a resolution of less than 4 meters and they have the potential for very detailed impact detection. In the Arctic there have been few studies where IKONOS-2 images have been used to study lichen pastures of reindeer (Allard 2003; Kumpula 2006).

There is a critical need to minimize conflicts between arctic grazing systems and industrial development, particularly in Russia which remains in a state of drastic and rapid transformation (Klein 2000; Krupnik 2000). There is strong evidence that these impacts are reaching a critical threshold in YNAO, where the withdrawal of lands for industrial infrastructure has pushed a steadily increasing number of animals onto progressively smaller areas of tundra pastures (Zen'ko 2004; Stammler 2005). This is leading to desertification in places and a serious decline in the quantity and quality of the remaining upland tundra suitable for reindeer pasture (Forbes 1999; Jernsletten & Klokov 2002). At the same time, the indigenous Nenets population has been exposed to a number of new health and demographic problems directly related to the industrial development (Pika & Bogoyavlensky 1995).

In the project *Environmetal and Social Impacts of Industrial Development in Northern Russia* (ENSINOR) we focus on a multidisciplinary analysis of the social and environmental consequences of energy development. We combine state-of-the-art Western scientific technologies with the indigenous or local perceptions of processes of change by Nenets reindeer herders. This method of co-producing knowledge can be expected to yield results which are both highly relevant to local and regional needs and practical for policy implementation. The immediate objectives are to characterize the effects of 2-3 decades of industrial development of petroleum on traditional livelihoods and the supporting society in YNAO and NAO. In this article we are studying the impacts that can be detected from the different resolution satellite images. We also discuss the significance of field surveys and local knowledge in enhancing the interpretation of satellite images.

Research area

The Yamal-Nenets Autonomous Okrug (YNAO) belongs to the West Siberian economic region and the Ural Federal district. YNAO is an independent unit of the Tyumen Oblast and lies in the extreme north of the West Siberian lowland, which encompasses the largest wetland (technically a peatland) in the world. With a territory of 750 300 km^2, the length of YNAO from north to south is 1200 km and from east to west 1130 km. In the Nenets' own language, "Yamal" means "the end of the earth" or "land's end". Ice-rich permafrost is widespread and is susceptible to thermokarst erosion from both natural and anthropogenic disturbance. As a result, much of the terrain is considered moderately to extremely unstable for purposes of engineering and infrastructure developments (roads, bridges, pipelines, etc.). The forest-tundra transition zone covers a broad band coincident with the southern portions of Yamal and Taz Peninsulas and most maps depict the latitudinal treeline as running approximately through the city of Salekhard on the Arctic Circle (lat. 66°33'N). For the indigenous Nenets, their needs are met mainly by reindeer. South of the Ob River delta, sparse lichen woodlands of Siberian larch (Larix sibirica) give way to more or less closed forest in the southernmost parts of YNAO. On the shores of lakes, rivers and bays, subsistence fishing is of major importance too. Substantial populations of terrestrial wildlife still exist, although some fur-bearing species are subject to hunting and trapping, both licit and illicit.

Today the YNAO is numerically the world's most productive reindeer herding region with 556 000 domestic reindeer, herded by approximately 13 000 nomadic Nenets and to a lesser extent Komi and Khanty families. While successor enterprises of Soviet state farms, sovkhoz dominate the institutional landscape in reindeer herding, private clan communities dynamically develop in the North of the okrug and hold more than half of the reindeer population. At the same time, the gas deposits discovered here are the largest worldwide. Large-scale industrialisation since the 1970's has led to an influx of people from the south, which is why today the indigenous share of the overall population is less than 7%.

This paper focuses on the vast Bovanenkovo gas field in the northwestern part of the Yamal Peninsula (Fig. 1). The gas field is located in the area to which the Yarsalinski reindeer sovkhoz holds the principal land title. Bovanenkovo is on the migration path of two major Yarsalinski reindeer brigades. These brigades reach Bovanenkovo from south in early-mid July on the way to the Kara Sea coast where reindeer are taken to for insect relief and high quality forage. Brigades return through Bovanenkovo in mid-late August when they start migration towards winter pastures on the south side of the Ob river. In addition, at least a dozen private reindeer herding camps use the larger Bovanenkovo area as reindeer grazing grounds. In comparison to the brigades, their herds are smaller in size, their migration routes shorter, and their presence in Bovanenkovo area longer, reaching from early summer until October-November. Very little is known about the exact migration patterns of these private herders. However, their number has been increasing after the fall of the Soviet Union, and their herds as well as their families probably outnumber those of the officially registered herding brigades today.

Materials and methods

Field surveys

Two field expeditions to Bovanenkovo gas field and to its surroundings were made

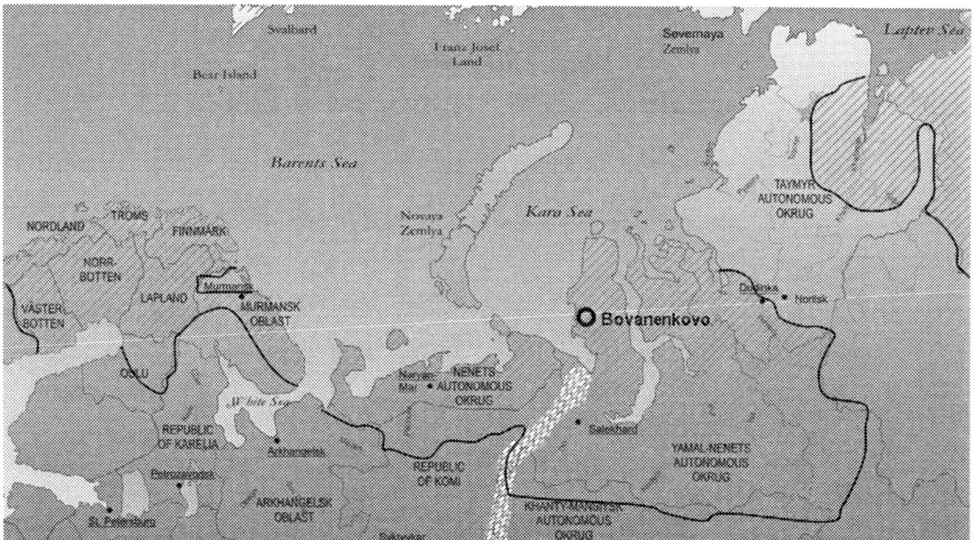

Figure 1. Reindeer herding in northern Fennoscandia and Siberia. Bovanenkovo gas field in Yamal Peninsula is marked with a black circle (lat. 70° 20'N long. 68°30'E). (Map: Arctic Centre, University of Lapland)

during the summers 2004 and 2005. A few hundred field sites were surveyed for image classification. Vegetation types were defined and special attention was placed on the signs of human impacts. Brigades migrating through the area were interviewed and also gas workers from Bovanenkovo were interviewed. Valuable information about Bovanenkovo's history and exploration were gained from both reindeer herders and gas workers. Herders from two brigades were accompanied during their migration through the gas deposit, and explained the migration history, their relations with the gas workers, changes and especially effects of gas exploration on everyday reindeer herding and reindeer pastures. Printouts of satellite images and topographical maps were used in interviews. Information from herders was also obtained during participant observation and in depth unstructured interviews that provided deeper insights about the relations of herders to the land, including its spiritual significance and the impact that the presence of industry has on these relations.

Remote sensing and GIS data and methods

Satellite images with different spatial resolution were used to compare the visibility of the different impacts. The very high resolution remote sensing data used in this study was Quickbird-2 (dated 15.7.2004). The spatial resolution of Quickbird-2 multispectral channels is 2.5 meters and on the panchromatic channel it is 0.61 cm. Quickbird-2 has 4 multispectral channels of blue, green, red and infrared. Data are stored in 11 byte format. High resolution data used were ASTER TERRA with 15 meter resolution (4 channels) (dated 21.7.2001), Landsat TM with 30 meter resolution (7 channels) (dated 07.08.1988), and Landsat MSS with 80 meter resolution (4 channels) (dated 28.7.1984) (Table 1).

GIS data with adequate accuracy were not available. Road network, sand quarries and drilling sites were digitized using Quickbird-2 and ASTER TERRA images. Visibility of different impacts on satellite images was compared between the images using field sites, ground photographs and photographs taken from helicopter. Reindeer herders' migration routes and brigadiers' notes were digitized from satellite image printouts. The aim was to produce a map depicting the total area of disturbance around the Bovanenkovo. Impact area estimation was produced by using field sites, photographs taken from helicopter and by visual interpretation of ASTER TERRA and Quickbird-2 images. Estimation was carried out by outlining the furthest visible signs of industrial impacts from around Bovanenkovo gas field. The area outlined in the ASTER TERRA image encompasses all visible signs of disturbance, such as roads, off-road vehicle tracks, quarries, etc.

Results

Detecting the impacts

From the remote sensing point of view it is always easier to detect the surface characteristics when trees and their canopy and shadows are absent. The detection of impacts from both the satellite data used

Table 1. Spectral characteristics of the satellite data used.

Spectral band	Quickbird-2 Wavelength (µm)	ASTER TERRA Wavelength (µm)	Landsat TM Wavelength (µm)	Landsat MSS Wavelength (µm)
1 (blue)	0.45 - 0.52	-	0.45 - 0.52	0.5 - 0.6
2 (green)	0.52 - 0.60	0.52 - 0.63	0.52 - 0.60	0.6 - 0.7
3 (red)	0.63 - 0.69	0.63 - 0.69	0.63 - 0.69	0.7 - 0.8
4 (NIR)	0.76 - 0.89	0.76 - 0.86	0.76 - 0.90	0.8 - 1.1
5 (NIR)	-	-	1.55 - 1.75	-
6 (TIR)	-	-	10.40 - 12.50	-
7 (NIR)	-	-	2.08 - 2.35	-
Panchromatic	0.45 - 0.90	-	-	-

and the field surveys is presented in Table 2. Contamination of the soil because of oil and various chemicals is difficult to detect even in the field if the contaminated area is small or if accidents have occurred some time ago and have since revegetated. Only a few possible sites were found where there were some sign of oil spills but which must have been leaked from a vehicle or an oil barrel (Bovanenkovo is a gas field). In satellite images there were no visible signs of oil or chemical contamination.

The removal of top soil and vegetation takes place in areas that have been used as building sites, vehicle storage areas, drilling sites activity etc. They usually extend over some acres or hectares and can be detected even from coarse resolution Landsat TM images. Larger quarries where sand is mined for road and other building purposes are detectable also from Landsat MSS images. This is because the spectral reflectance of bare ground is significantly different compared to the reflectance of vegetation.

Garbage is unfortunately difficult to detect. In the field there can be piles of leftover wood and metal gathered together that is possible to detect from panchromatic Quickbird-2 imagery. Wood remaining on abandoned drilling sites is of high value to reindeer herders who have to cope with a scarcity of wood for cooking and other purposes. Herders also make extensive innovative use of other leftovers on former industrial facilities and by doing so contribute to the recycling of waste to a certain extent (Stammler 2002). When metal and glass waste is sparsely distributed,

Table 2. Capacity to detect different impacts of petroleum exploration in Bovanenkovo. Scale is –not visible, x visible with effort, xx quite visible, xxx clearly visible.

Capacity to detect Impact	**Field survey**	**Quickbird-2 Panchromatic**	**Quickbird-2 Multispectral**	**ASTER TERRA VNIR**	**Landsat TM**	**Landsat MSS**
Soil contamination, oil & chemicals	x	-	-	-	-	-
Removal of top soil and vegetation	xxx	xxx	xxx	xx	x	x
Quarries	xxx	xxx	xxx	xxx	xx	x
Garbage						
- metal	xx	-	-	-	-	-
- glass	x	-	-	-	-	-
- concrete	xxx	x	x	-	-	-
- wood	xxx	x	-	-	-	-
Pipelines	xxx	xx	x	-	-	-
Power lines	xxx	xx	x	-	-	-
Roads	xxx	xxx	xxx	xxx	x	x
Off-road vehicle tracks	xx	xxx	xx	xx	x	x
Winter roads	xx	xx	xx	xx	x	-
Drill towers	xxx	xxx	xx	x	-	-
Barracks	xxx	xxx	xx	x	-	-
Trucks/Vehicles	xxx	xx	x	-	-	-

detecting them in field survey can be difficult. Garbage can partly or completely be covered with vegetation. This kind of garbage is one of the most harmful to reindeer according to herders. Reindeer can injure their hooves from glass or sharp metal and hence be exposed to infections that may in the worst scenario lead to death.

Pipelines and power lines are reasonably detectable in multispectral Quickbird-2 imagery. Roads, winter roads and off-road tracks are just barely visible in Landsat MSS image. Drill towers and barracks can be detected with ASTER TERRA imagery. Vehicles can also be seen from multispectral Quickbird images.

Field survey was the most useful way to identify different impacts. Some objects like revegetated old tracks, some seldom used off-road tracks and areas that have been revegetated completely were easier to detect and outline from multispectral Quickbird-2 image. Noticeable is that some tracks were impossible to detect in the field before we checked the satellite image printout.

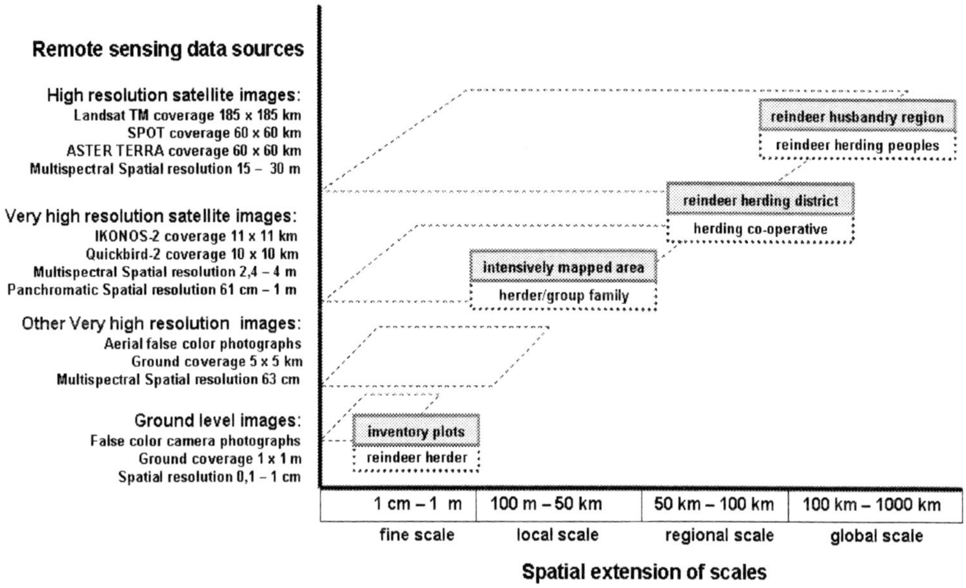

Figure 2. Scales of investigations and relevant remote sensing data sources (Timo Kumpula & Benjamin Burkhard).

Scales of investigations

When investigating environmental impacts, the data resolution has to be suitable for research purposes. For example pipelines cannot be detected from Landsat TM images accurately. Size of the research area is another factor determining the data to be used either in terms of data size (megabytes) or data costs. In the case of Bovanenkovo, where the total area is 35 * 10 km, ASTER TERRA is the most cost efficient data source to study impacts of the whole gas field. But if more detailed study is planned, Quicbird-2 image is a very good choice. Quickbird-2 multispectral image enhanced with panchromatic image provide the best data to detect impacts around drilling sites. Most impacts are detectable in this image combination (Fig. 2).

ASTER TERRA images are worthy of attention from among the larger scale sources of satellite data. Landsat TM covers a larger area and has better spectral resolution with 7 channels, but ASTER TERRA's 15 meter resolution and lower price increases its overall utility to detect environmental impacts.

Area of disturbance

Total area of disturbance was calculated by outlining inside Bovanenkovo the areas that had signs of human disturbances visible in ASTER TERRA and Quickbird-2 images (Fig. 3). The area of disturbance covers 317 km^2. Inside this buffer there are visible changes compared to the original vegetation cover. Inside this area there is also a significant amount of relatively undisturbed vegetation. Permanently lost pasture-land includes roads and infrastructure areas. The active road network totals 79 km and it covers 143 ha, by calculating an 18 m wide buffer around road network. Although roads are only approximately 18 m wide, the actual affected zone around them is much broader due to the cumulative impacts of blowing sand and dust and altered hydrology. Significant changes in soil moisture, pH and plant chemistry are detectable at least 35 m from the edges of roads within four years of their construction (Forbes 1995). Infrastructure (drill sites, barracks, storage areas) covers 203 ha and quarries 373 ha. However, we are aware that besides these detectable impacts on satellite images, reindeer herders are heavily impacted by the social and economic consequences of industrial infrastructure: for example, a road opens up larger areas of the tundra for open access by industrial workers. Besides for working purposes, they may use roads for going on hunting, fishing (poaching), recreational trips or trading. Being untrained for the interaction with herders and herds, they might not behave in an appropriate way in the vicinity of herders and herds. Therefore, roads are significant for herders beyond the immediate corridor of impacted pasture.

Herders' and workers' interpretation of satellite images

Interviewing reindeer herders was essential to understanding the impacts of petroleum industry on reindeer and their herders. Information on the history of exploration

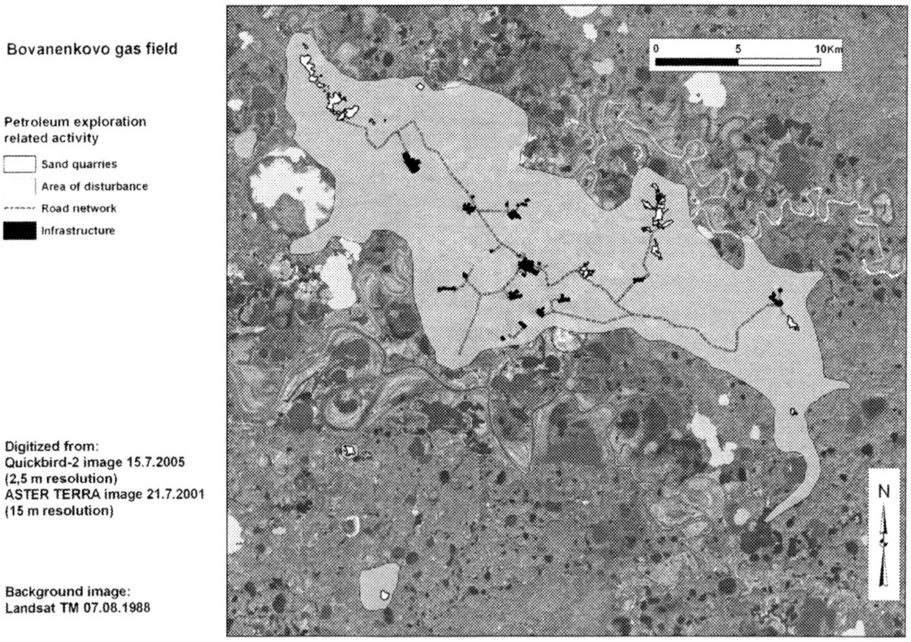

Figure 3. Area of disturbance around Bovanenkovo gas field.

which has taken place in the area would have been extremely difficult or totally impossible to gain from other sources. Many herders were able to interpret satellite image printouts extremely efficiently (Fig. 4). False colour composites did not cause problems to identify lakes, rivers, campsites and migration routes with their exact locations. Herders through their perception of the migration route and the industrial area added an important level of historical depth and detail to our understanding of recent changes in the area. Gas workers also provided a lot of information about the area and history from their point of view. They also provided information concerning the future of the petroleum exploration in the area. Interviews and participant observation with both groups of workers in the tundra impressively revealed the various mental maps of people with different interests in the area. The mental map of most herders consists of a complex set of migration routes, pastures, fishing grounds, camp sites presenting a detailed image of the on-surface resources. The gas workers' mental map in the Bovanenkovo area was also very detailed and in many cases reached back decades, consisting of a complex set of drilling sites that provide access to different layers of the gas under the permafrost, and of pipelines, roads, and camps dispersed across the whole territory of the gas deposit. Gas workers also had developed throughout the decades an intimate knowledge of the rivers and lakes, some of which they have been using for fishing for a long time, and some of which are now being heavily impacted by new road constructions leading right through the water bodies.

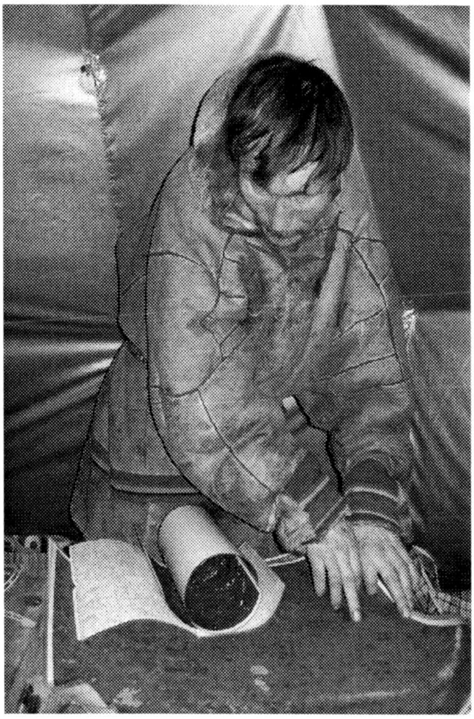

Figure 4. Brigadier Vasili Serotetto explaining migration routes, campsites and pastures with false-colour composite of ASTER TERRA imagery. Photo: Timo Kumpula 14.7.2005.

The technology of building bridges by blocking off small rivers or lakes completely for the duration of the road works has been heavily criticised by herders. As a result, larger areas around these blocked rivers can become flooded and turn into extreme wetlands that are not suitable as reindeer pasture or migration routes anymore. These flooded areas as well as drained lakes can also be detected on satellite pictures and herders added historical depth and detailed descriptions of events for these areas.

Discussion

The preliminary results in detecting the impacts of oil petroleum exploration on reindeer herding were promising. It seems that combining different scale remote sensing data results in impacts detection can be achieved. Usually drillings sites are spatially quite small and like Bovanenkovo the whole gas field is 35 * 15 km. This allows pinpointing the special area of interest and to acquire expensive very high resolution data like Quickbird-2 or IKONOS-2. Then detailed research on impacts is possible to conduct.

During the field survey the limitations of human eye were once more clearly demonstrated. We noticed that in the field it was difficult to notice revegetated vehicle tracks and the edges between once disturbed and untouched vegetation. In satellite image printouts different intensities of infra-red signatures were very clear. Also detecting old vehicle tracks from ground level was much more difficult than from helicopter and Quickbird-2 images.

Local and indigenous peoples' assessment was crucial in understanding current development and impacts of petroleum exploration. In general, they add a different level of interpretation to changes on the tundra by ranking events, development and changes and evaluating them in terms of their positive or negative impact on their livelihoods. Thus, satellite images and biological ground truthing provide an important factual basis and information about their significance for the lives of humans and animals on the tundra. Herders are able to provide qualitative data (testimony from active and retired herders) on the degree of environmental and socio-economic change during past decades. They provide information on the current variation in the vegetation and reindeer habitat types and how they are ranked according to quality. They can also estimate what proportion of the habitat has been lost because of the petroleum industry and how the herders and reindeer have adapted to this. Herders also have a good idea where areas of potential or likely future degradation may occur (e.g. desertification from heavy grazing, melting of frozen ground). Herders' expectations or scenarios for the future under petroleum development are important factors that should be considered in research question development. Nenets developed strategies for mitigation of predicted negative changes and scenarios of future reindeer management in conjunction with petroleum development.

What are more difficult or impossible to detect on satellite images are the potential and actual positive social and economic impacts of industrial development on reindeer herders. During fieldwork, participant observation with them revealed that in principle the presence of industrial workers and their facilities is welcomed by herders of the area, and their perception of the negative effects is balanced by the benefits, for example of the workers' settlements as a source of staple food to resupply, the use of roads for the supply of herders, the settlements as a market for reindeer meat, the helicopters of the gas company as means of transport and as means to collect velvet reindeer antlers to be sold on far eastern markets for cash (Stammler 2004). In the case of the private herding camps not belonging to officially registered reindeer brigades, the gas company has become the most important trading partner.

The further processing of satellite data focuses on the extent and rate of tundra ecosystem change during the past 20 – 30+ years. Landsat MSS and TM are the main data sources when studying the situation in the 1970's and 1980's but also CORONA spy satellite images can be used in change detection. The aim is to estimate how large a fraction of the habitat has been affected within the current range of reindeer habitats. GIS and risk assessment of potential future degradation in expanding drilling sites will be conducted.

Together the applied results from these assessments will provide a potentially powerful tool for regional planning and policy development. The results will be of interest to scientists in several disciplines (e.g. anthropology, geography, biology/ecology, political science, international relations, conservation biology), as well as to indigenous groups elsewhere in other arctic and subarctic regions.

Acknowledgements

We would like to thank the Finnish Academy for financing the project Environmetal and Social Impacts of Industrial Development in Northern Russia (ENSINOR) 2004-2007.

References

Allard, A. (2003). Detection of vegetation degradation on Swedish mountainous heaths at an early stage by image interpretation. *Ambio* vol. 32:8, 510-519.

Benelli, G. & A. Garzelli (1999). Oil-spills detection in SAR images by fractal dimension estimation, *Proc. IGARSS'99* vol. 1, 218–220.

Espedal, H.A. & O.M. Johannessen (2000). Detection of oil spills near offshore installations using synthetic aperture radar (SAR). *International Journal of Remote Sensing* vol. 21: 11, 2141–2144.

Forbes, B.C. (1995). Tundra disturbance studies III. Short-term effects of aeolian sand and dust, Yamal Region, northwest Siberia, Russia. *Environmental Conservation* vol. 22, 335-344.

Forbes, B.C. (1999). Reindeer herding and petroleum development on Poluostrov Yamal: sustainable or mutually incompatible uses? *Polar Record* vol. 35, 317-322.

Forbes, B.C. (2004). Impacts of energy development in polar regions. *In* C.J. Cleveland (ed.): *Encyclopedia of Energy*, 93-105. Academic Press, San Diego.

Ingold, T. (2005). A manifesto for the anthropology of the North. *In* Sudkamp A. (ed.): *Connections: local and global aspects of Arctic social systems*, 61-71. University of Alaska, Fairbanks.

Jernsletten, J.-L. & K. Klokov (2002). *Sustainable reindeer husbandry*. Arctic Council/Centre for Saami Studies, Tromsø.

Jones, B.A. (2001). Comparison of visual observations of surface oil with synthetic aperture radar imagery of the Sea Empress oil spill. *International Journal of Remote Sensing* vol. 22: 9, 1619–1638.

Klein, D.R. (2000). Arctic grazing systems and industrial development: can we minimize conflicts? *Polar Research* vol. 19, 91-98.

Krupnik, I. (2000). Reindeer pastoralism in modern Siberia: research and survival in the time of crash. *Polar Research* vol. 19, 49-56.

Kumpula, T. (2006). Very High Resolution Remote Sensing Data in Reindeer Pasture Inventory in Northern Fennoscandia. *In* Forbes, B.C., M. Bölter, L. Müller-Wille, J. Hukkinen, F. Müller, N. Gunslay & Y. Konstantinov (eds.): *Reindeer Management in northernmost Europe*, 167-185. Springer, Berlin.

Meschtyb, N.A., B.C. Forbes & P. Kankaanpää (2005). Social Impact Assessment along Russia's Northern Sea Route: petroleum transport and the Arctic Operational Platform (ARCOP). *Arctic InfoNorth* 58, 322-327.

Mosnews (2006). Russia Overtakes Saudi Arabia as World's Leading Oil Producer. 11.09.2006. *http://www.mosnews.com/money/2006/08/23/russiaoil.shtml.*

Pika, A. & D. Bogoyavlensky (1995). Yamal Peninsula: oil and gas development and problems of demography and health among indigenous populations. *Arctic Anthropology* vol. 32, 61-74.

Rees, W.G., M. Williams & P. Vitebsky (2003). Mapping land cover change in a reindeer herding area of the Russian Arctic using Landsat TM and ETM+ imagery and indigenous knowledge. *Remote Sensing of Environment* vol. 85: 4, 441-452.

Solberg, A.H.S., S.T. Dokken & R. Solberg (2003). Automatic detection of oil spills in Envisat, Radarsat and ERS SAR images. *Proc. IGARSS'03* vol. 4, 2747–2749.

Stammler, F. (2002). Success at the edge of the land: Present and past challenges for reindeer herders of the West-Siberian Yamal-Nenets autonomous Okrug. *Nomadic Peoples* NS vol. 6, 51-71.

Stammler, F. (2004). The Commoditisation of Reindeer Herding in Post Soviet Russia: Herders, Antlers and Traders in Yamal. *In* Leder S. & B. Streck (eds.): *Segmentation und Komplementaritaet. Organisatorische, oekonomische und kulturelle Aspekte der Interaktion von Nomaden und Sesshaften. Mitteilungen des SFB "Differenz und Integration"*. Orientwissenschaftliche Hefte vol. 14, 105-122.

Stammler, F. (2005). *Reindeer Nomads Meet the Market: Culture, Property and Globalisation at the End of the Land.* Halle Studies in the Anthropology of Eurasia vol. 6. Lit publishers, Münster.

Stammler, F. & E. Wilson (2006). Dialogue for development: an exploration of relations between oil and gas companies, communities and the state. *In* Wilson E. & F. Stammler (eds.): *Special issue on the oil and gas industry local communities and the state*, 1-42. Sibirica vol. 5: 2.

Stow, D.A., A. Hope, D. McGuire, D. Verbyla, J. Gamon, F. Huemmrich, S. Houston, C. Racine, M. Sturm, K. Tape, L. Hinzman, K. Yoshikawa, C. Tweedie, B. Noyle, C. Silapaswan, D. Douglas, B. Griffith, G. Jia, H. Epstein, D. Walker, S. Daeschner, A. Petersen, L.M Zhou & R. Myneni (2004). Remote sensing of vegetation and land-cover change in Arctic Tundra Ecosystems. *Remote Sensing of Environment* vol. 89, 281-308.

Toutoubalina, O.V. & W.G. Rees (1999). Remote sensing in detection of industrial impact around Noril'sk, northern Siberia: preliminary results. *International Journal of Remote Sensing* vol. 20: 15-16, 2979-2990.

Tømmervik, H., K.A. Høgda & I. Solheim (2003). Monitoring vegetation changes in Pasvik (Norway) and Pechenga in Kola peninsula (Russia) using multitemporal landsat MSS/TM data. *Remote sensing of Environment* vol. 85, 370–388.

Virtanen, T., K. Mikkola, E. Patova & A. Nikula (2002). Satellite image analysis of human caused changes in the tundra vegetation around the city of Vorkuta, north-European Russia. *Environmental Pollution* vol. 120: 3, 647-658.

Zen'ko, M. A. (2004). Contemporary Yamal: ethnoecological and ethnosocial problems. *Anthropology & Archaeology of Eurasia* vol. 42, 7-63.

The Forgotten Uranium Mine of Paukkajanvaara, North Karelia, Finland

Alfred Colpaert
Department of Geography, University of Joensuu

Abstract: The cold war era was a time with huge demand for uranium. Also in Finland exploration showed the presence of uranium in the bedrock of North Karelia. A small mine operated from 1959 to 1961 near Paukkajanvaara in the municipality of Eno. The mine produced some 30 tons of uranium concentrate. After the main ore body was exhausted the mine was abandoned, due to the low demand for uranium after 1960, the mine has never been reopened. The area was without attention for nearly three decades. During the 1990's the area was rehabilitated, covered with clay and till and replanted with pine. After the land rehabilitation project radiation in the mining site was reduced to the normal Finnish background level. The site is an interesting field laboratory to study the release of radioactive elements from uranium mining into the environment. Knowledge from this site could prove to be important, seen in the light of the renewed interest in uranium exploration in Finland.

The Paukkajanvaara uranium deposits

Uranium is an element found in very small quantities in the earths crust. Under some special geological conditions uranium can be enriched thousand fold into ore deposits and concentrations may be sufficient to start mining operations. Uranium became an important commodity after the Second World War when both the United States and the Soviet Union started to build up their nuclear arsenals. This and the need of uranium for nuclear power plants created a huge demand for uranium. Exploration for uranium resources became a lucrative business and many part-time prospectors invested money in simple Geiger counter equipment.

In 1957 two brothers, Eino and Martti Justander from Eno, North Karelia bought a Geiger counter and started exploring the area around their home near lake Herajärvi (Fig. 1). Soon they found a radioactive metabasite outcrop from which they sent samples to Atomienergia Oy, the Finnish firm responsible for atom energy in Finland at the time. Further field investigations by the geologist Piirainen revealed radioactive erratic boulders. Samples from these proved to contain 1.5 % uranium and this led to intensive exploration in the area (Piirainen 1968). The average concentration of uranium in the earth's crust is far less, about two to three mg/kg, which is 10.000 times less than the samples from Eno. The

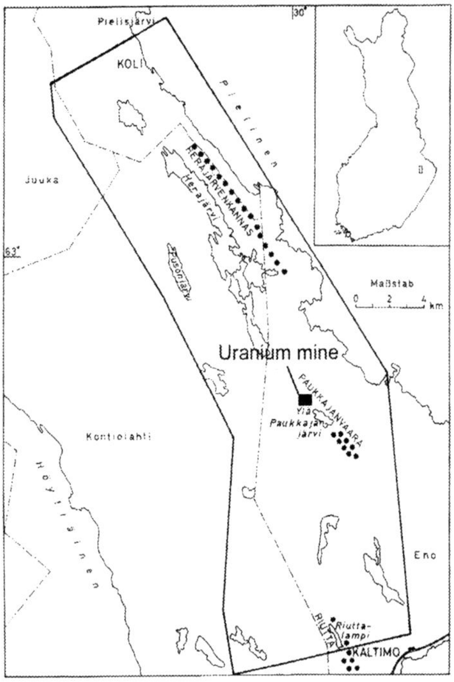

Figure 1. Location of the boulders containing uranium ore (black dots) and the uranium mine (from Piirainen 1968, modified by the author).

explorations were carried out swiftly and led soon to a mining claim. During 1957 – 1958 numerous radioactive erratic boulders were found in the Herajärvi – Riutta area. During the 1960's the area has been intensively explored and various small uranium deposits have been identified.

The radioactive boulders were deposited in a string-like fashion in NW-SE direction, as they have been plucked from the bedrock and transported during the last glaciation. The uranium is deposited in small local metadiabase dykes and metamorphic-sedimentary contacts with these intrusions (Fig. 2).

From exploration to mining

In 1959 a small mining operation was started at the northern shore of lake Ylä-Paukkajanjärvi, called the Paukkajanvaara mining site (Fig. 3). The mine was basically a test facility to assess the feasibility of large scale uranium mining. The site consisted of the Mårtenson quarry, which was soon transferred into an underground mine and close to this a second, smaller quarry. The ore was milled and enriched at the mining site. Several tens of workers were employed at the mine. No special protective gear was used but workers were however obliged to carry dosimeters (Tauno Piirainen, pers. comm.). Radiation caused by radon gas has been a major problem in many mines due to insufficient ventilation. At present dust and other normal mining hazards are probably more important health issues. Although radiation is higher near uranium bearing rock the dose is still low, for the Paukkajanvaara site it is in the range of 1 – 9 µS/h, the Finnish average being about 0.1 µS/h (Mustonen & al. 1989). From the Paukkajanvaara mining site there have been no reports of negative effects on the health of workers, which would of course be difficult to prove due to the limited number of personnel and short production period.

Between 1959 and 1961 about 30 700 tons of uranium ore was extracted from the mine, 21 400 from underground and 9 400 tons from the open pit mine. The uranium content was between 0.075 (quarry) and 0.143 % (Mårtenson mine) (Sillanpää & al. 1989). About 27 tons of uranium ore concentrate (20 – 30 % purity) was produced from the crude ore. The purified uranium ore was sold to Sweden. The Mårtenson

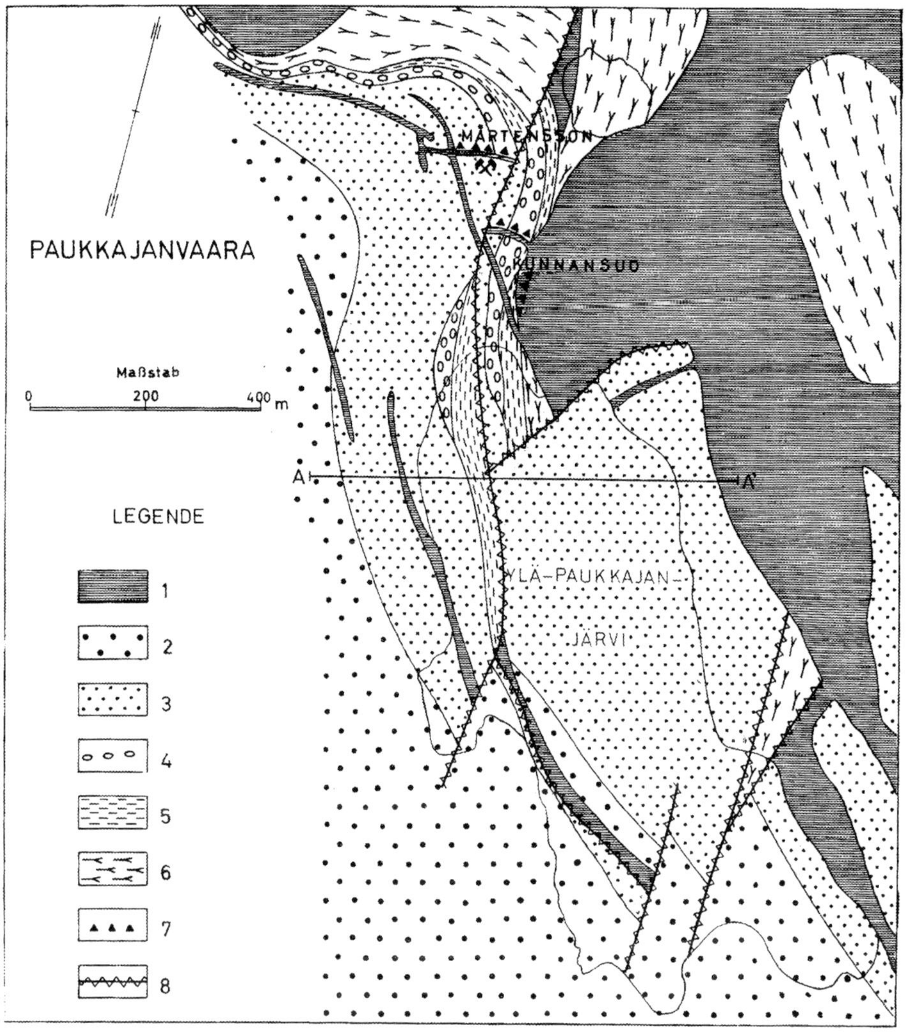

Figure 2. Geology of the Paukkajanvaara area (Piirainen 1968).1 = metadiabase, 2 = arkosic quartzite, 3 = ortho quartzite, 4 = conglomerate, 5 = sericite quartzite, 6 = granite, gneiss 7 = uranium ore, 8 = fault (Piirainen 1968, published with permission from the author).

uranium ore was exhausted already in 1961 and mining of the Kunnansuo deposit was not economically profitable. After the operations stopped in 1961 the site was abandoned. In 1974 the last purified uranium ore, about 2 300 kg which was left in a storage building was removed and transported to the facilities of Outokumpu Oy in Pori. The production buildings were demolished between 1976 and 1980, the entrance of the mining shaft was covered with a concrete slab, the rest of the area, tailings, waste-water ponds etc. were left as they were. In 1985 some earth moving was done to uncover

Figure 3. The Mårtenson uranium mine in 1960. Photo: Ilkka Laitakari, Geological Survey of Finland.

mica containing rock strata and a new 200 hectares mining claim was made containing the old uranium mining site. The site remained dormant and no production was ever resumed.

Uranium and radiation in the environment

After the closure of the mine, it was forgotten by all but the local residents and media. In 1962 the local water authorities made an inquiry about the safety of fish in the area, the response of the local authorities was short and clear, no contaminated materials could be released into the downstream area (Letter to the... 10.11.1962). The Finnish Centre for Radiation and Nuclear Safety started a major study, monitoring air, water and soil in and around the site. The area had been untouched for nearly 30 year and was therefore an ideal study site for dispersal of radioactive material into the environment. Previously measurements had been made in 1974 and 1979, but in 1984 systematic sampling of the area and a geologically similar undisturbed reference area (Riutta) started. The project ended in 1989. The radiation at the mining site was estimated to have been increased 0.5 µS/h as a direct result of the mining operations. Radon levels were exceptionally high in the air above the mining site, this was mainly caused by the tailings consisting of fine grained material from the milling and leaching process. About 12 000 m^3 of fine grained material and about 7 300 m^3 of waste rock were present in the tailings. On top of the tailings measured values were 150 – 300 Bq/m^3, for the whole site a value of 100 Bq/m^3 was assumed (Sillanpää & al. 1989). The Finnish maximum level for indoor values has been set to 400 Bq/m^3 (STUK 2006).

Contamination of surface water occurred as a small ditch was flowing through the mining site transporting sediments towards the Iso-Hiisilampi pond (about 200 meter downstream) from where it flowed into the stream draining into lake Pielinen. Radiation levels in the surface water were considerably higher than in the Riutta reference area, 0.2 – 0.4 Bq/l compared to 0.002 – 0.003 Bq/l (Sillanpää & al. 1989). Radiation level of the surface water was normal about two kilometer downstream were it flows through a small lake (Saarilampi) and becomes more diluted. The amount of dissolved uranium in the surface water downstream from the mine was about 20 to 25 $\mu g/l$, values decreased slowly reaching normal levels when entering lake Pielinen. Values for the Riutta reference area were much lower, 0.2 to 0.5 $\mu g/l$ (Sillanpää & al. 1989).

Radiation values from uranium for bottom sediment were much higher, 10 000 to 30 000 Bq/kg (dry mass) for the bottom of the pond just downstream of the mining site, reducing sharply to 1 000 Bq/kg and becoming normal some two kilometers downstream in the Saarilampi lake, for the reference area values were 100 – 400 Bq/kg.

Plants and mushrooms had only slightly higher uranium contents, and this only in those plants growing directly on the mining site. Fish, however, had 20 to 30 fold uranium levels as compared to the Riutta reference area. The activity was highest in fish for the direct downstream area but also fish form the Saarilampi lake had relatively high amounts of radioactive isotopes (Mustonen & al. 1989).

The conclusions by Mustonen & al. (1989) are that rehabilitation was needed if the radiation levels of the mining site were to be ameliorated to the original levels. They also conclude that the radiological effects of the uranium mining activity could be traced all the way downstream to lake Pielinen.

Rehabilitation of the mine

In 1991 planning of the rehabilitation of the mining area started (Imatran Voima 13. 12.1991). The total cost of removing the contaminated sediments and veering of the drainage were estimated to be 680 000 Finnish marks (about 113 000 €). The project started in 1992 and was finished in 1994. Most of the planned actions were implemented, but the small ponds and stream sediments were not removed. The Mårtenson mine shaft and pit were demolished with dynamite. The whole area including tailings was covered with 30 centimeter of clay on top of which 1.2 meter of till. The resulting area was landscaped into a gently sloping grade and finally pine trees were planted.

The final inspection of the area was done 23.08.1994 and the area was proclaimed to be without restrictions for outdoor use. The building of houses and cottages however remained prohibited. Measurements by the Finnish Centre for Radiation and Nuclear Safety were made and showed only near the previous enrichment station slightly above average values, all other parts of the

Figure 4. Old sign at the entrance gate with the text "Radioactivity, access prohibited". Photo: Alfred Colpaert 18.8.2004.

area were at the normal Finnish average (Bonvesta 23.8.1994).

When we visited the area in 2005 there were only few signs left of the former mining area, only a rusted sign at the gate (Fig. 4) reminds us of the former history.

The rehabilitated area itself looks like any planted pine forest (Fig. 5), some quarry areas can still be distinguished. The area is still drained by a small ditch running through the middle of the area. We did some radiation measurements with a DGM meter from Kata Oy, and did not find signs of radiation above the normal base level (0.08 μS/h on our device). Only in the gravel near the entrance of the area slightly higher values were found 1.32 μS/h (Fig. 6).

Renewed interest in uranium exploration

Uranium exploration and mining was not very profitable during the last decades of the previous century due to environmental issues related to atomic energy. Countries like Sweden and Germany were de facto stopping production of atomic energy, and in many countries plans for new reactors were put on hold. Also the availability of weapons grade nuclear material which was diluted and sold for peaceful energy

Figure 5. The rehabilitated mining area. Photo: Alfred Colpaert 18.8.2004.

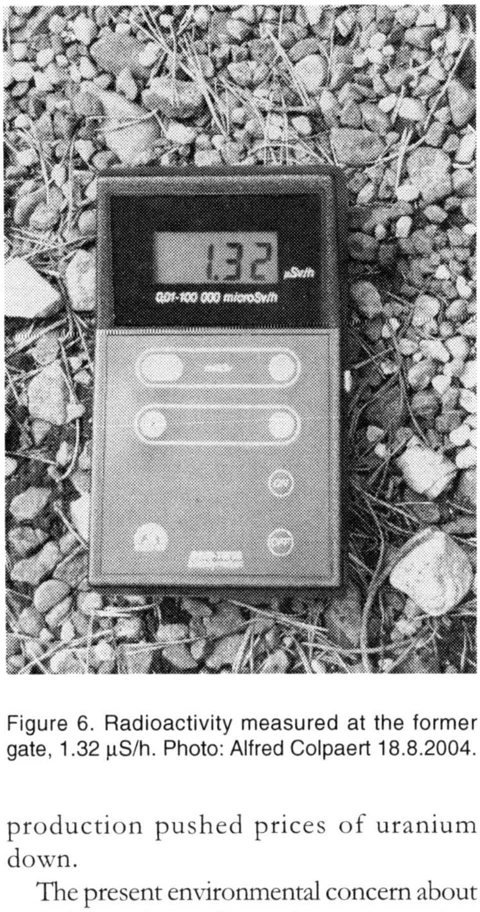

Figure 6. Radioactivity measured at the former gate, 1.32 µS/h. Photo: Alfred Colpaert 18.8.2004.

production pushed prices of uranium down.

The present environmental concern about global warming and greenhouse gasses has renewed the interest in nuclear power production. Trading in carbon dioxide has made nuclear energy more profitable and former stock piles of decommissioned nuclear weapons are exhausted. Also the rise in price of fossil fuels, like oil and gas and the dependence on these has led many countries to review their stand on nuclear energy.

As uranium is commonly found in areas like Russia, Canada and Australia, areas with old Archean age bedrock with granite intrusions and contact areas, exploration in Finland has been intense during the last years. The whole eastern region along the Russian – Finnish border from Porvoo in the south to Inari in the very north is under investigation. Some claims have already been made, near Porvoo and Eno by the French mining company Cogena, and by the Finnish company Namura in Kuusamo (KTM 2006).

Conclusions

The former uranium mine Paukkajanvaara near Eno was by any standards a very small operation. It operated effectively only for three years and produced a limited amount of waste rock. It was easy to rehabilitate and presently there is no measurable radiation at the site, only the bottom sediments of the nearby pond and stream could possibly contain some traces of uranium. In respect to possible large scale uranium mining in Finland the site could prove to be a valuable study object. For example the fact that a small stream was allowed to flow through the mining site was a mistake not taken into account during the 1960's. At the time it was believed that due to its large specific weight uranium would not leave the site by air or water. Although uranium content of rock and ore is very low and direct radiation is relatively small, the mining, milling and tailings produce radon gasses which can be a health problem.

The lesson from the Paukkajanvaara mine is that direct and short-term effects of uranium mining are not very great, the main problem is the long lasting effects of waste material leaching and water and sediment contamination.

Well managed and maintained, under strict environmental and radiological control uranium mining could be safe. It must however be considered as special and the highest caution and safety standards for both workers and the environment are needed. Rehabilitation should take place directly after ending of production, and mines should not be allowed to lay dormant. When a mine lies unattended for several decades the protection measures around tailings and other facilities tend to become unstable and leaching by rain water creates a possibility for ground- and stream water contamination.

Acknowledgements

I would like to thank emeritus professor Dr. Tauno Piirainen, former mine geologist Dr. Matti Tyni for their valuable information and comments and Dr. Heikki Vesajoki, who accompanied me on the mining site, for his assistance and comments.

References

Karhu, E. (2006). Uraani kiinnostaa kannattajia ja kiihottaa vastustajia. *Metallitekniikka* 2006: 6, 30-33.

Mustonen, R., T. K. Ikäheimonen, L. Salonen & T. Sillanpää (1989). Uraanin louhinnan ja rikastuksen radiologiset ympäristövaikutukset Enon Paukkajanvaarassa. *STUK-B-VALO* 61. Finnish Centre for Radiation and Nuclear Safety.

Piirainen, T. (1968). Die Petrologie und die Uranlagerstätten des Koli-Kaltimogebiets im Finnischen Nordkarelien. *Bull. Comm. Geol. Finlande* 237. Geological Survey of Finland, Helsinki.

Sillanpää, T., T. K. Ikäheimonen, L. Salonen, T. Taipale & R. Mustonen (1989). Paukkajanvaaran vanhan uraanikaivos- ja rikastamoalueen ja sen ympäristön radioaktiivisuustutkimukset. *STUK-B-VALO* 56. Finnish Centre for Radiation and Nuclear Safety.

WWW-material

KTM (2006). Ministry of Trade and Industry. *http://www.ktm.fi/index.phtml?s=1585*

STUK (2006). Finnish Centre for Radiation and Nuclear Safety *http://www.stuk.fi/sateilytietoa/sateily_ymparistossa/radon/fi_FI/radon/_files/71375051672125691/default/onko_ kodissasi _radonia.pdf*

Archive material

Letter to the local water authorities in the Paukkajanvaara area (10.11.1962). Archive of the Environmental Centre of North Karelia, Joensuu.

Imatran Voima (13.12.1991). Plan and cost calculation for rehabilitation of the Paukkajanvaaran mining area. Archive of the Environmental Centre of North Karelia, Joensuu

Bonvesta (23.08.1994). Minutes of the final inspection of the Paukkajanvaaran rehabilitation. Archive of the Environmental Centre of North Karelia, Joensuu.

Location-based Services as a Tool for Developing Tourism in Marginal Regions

Harri Antikainen[1*], Jarmo Rusanen[1], Sami Vartiainen[1], Mauri Myllyaho[2], Jari Karvonen[2], Markku Oivo[2], Jouni Similä[2] & Kari Laine[3]
[1] Department of Geography, University of Oulu
[2] Department of Information Processing Science, University of Oulu
[3] Thule Institute, University of Oulu
[*] Corresponding author

Abstract: Tourism is a key business sector for many geographically and economically marginal regions. The main form of tourism in these regions is nature-related tourism, consisting of activities such as hiking, fishing and various types of winter sports. The tourism resorts in marginal regions face the constant challenge of improving their tourist product to retain their position in the market while attempting to conserve the natural environment. In addition to the quest for pure excitement and adventure, the tourists of today expect better accessibility to information, services and nature. In response to this demand, a more widespread use of new information communication technologies, including mobile location-based services (LBS), is proposed in this paper. LBS have already been introduced to the tourism market, but mostly in urban areas. Adverse field conditions, inadequate availability of relevant data sources and high development costs are among the most common obstacles to the adoption of LBS in marginal regions. However, we conclude that LBS have the potential to add significant value to local tourism products and even create new business activities.

Introduction

Tourism differs from most other fields of business in the sense that it can be carried on in geographically and economically marginal regions. This is especially true of Finland and other Nordic countries, where the economic and political significance of tourism for regional development has increased considerably in recent years, to the extent that in some regions it has become a crucial economic policy factor. In these regions tourism essentially forms the main business on which other business sectors depend. The tourism industry is commonly looked on as having a favourable impact on economic and social development, introducing new, external sources of monetary income and opportunities to work in the regional economy and its operational cycle. For this reason, large public and private investments are being targeted nowadays at the development of tourism in the various regions of Finland (Butler & al. 1998; Saarinen 2003).

In the most marginalized regions in Finland, the natural environment, tranquillity and unique culture are considered to be among the most important tourist attractions, which offer the possibility for leisure-time activities such as hiking and fishing, and winter sports such as skiing and snowmobile safaris. As tourism is a demand-driven industry, tourism destinations have to constantly develop and improve their services in order to provide value to the tourist and retain the competitive advantage over other destinations. In nature areas, however, this involves a trade-off between more efficient use and access to the environment and the conservation of its natural condition and attractiveness. On the one hand, tourism consumes landscape, while on the other hand, it needs unspoilt nature to preserve its attractiveness (IFN Bulletin 2001).

As one of the possible tools to address the problem of the trade-off, the more widespread utilization of new information communication technologies has been proposed. The rapidly evolving technologies, especially mobile technologies with positioning capability, facilitate the development of new applications and services for travellers. They are expected to provide a means to respond to some of the emerging trends concerning nature tourism to marginal regions, such as the demand for better access to information and services, the need for improved guidance and safety, and generally increased environmental awareness of the tourists (Aichholzer & al. 2003).

This article is organized as follows. The first section covers some of the basic aspects pertaining to mobile location-based services and tourism. This is followed by the introduction of a research project that aims to study the topic presented in this paper, as well as a short and non-comprehensive review of related research efforts under way. The next two sections deal with different business models and implementation challenges with the emphasis on technological constraints. Finally, the possible implications of location-based services on tourism and regional development are discussed in the concluding section.

Location-based services for tourism

The widely adopted term location-based services (LBS) refers to information services that are accessible through a mobile handset and based on the current geographic location of the mobile device. The geographic location of the device is determined by using a separate positioning service, nowadays most commonly involving the use of the satellite-based Global Positioning System (GPS). The conventional application areas of LBS include mapping, tracking, routing and logistics, electronic yellow pages, data collection and public safety (Beaulieu & Cooper 2001; Maguire 2001; Veijalainen & al. 2001; Zipf & Malaka 2001). The common denominator of these applications is the use of a positioning capability as a value-adding factor.

The mobile technology is maturing and spreading and nowadays such enabling technology is widely available. New mobile terminals are enhanced with multimedia capabilities, faster data transfer, cameras, GPS, wireless links, etc., and their share of the market is increasing rapidly. Despite the popularization of the technology, it is not yet being efficiently utilized, and the related business models are in their early stages. However, tourism is among the first industry taking advantage of LBS. The main reason for this is that tourism can intrinsically benefit from the use of mobile technology, that provides new services to travellers on the move. In fact, the supply of mobile services with specific relevance to the traveller has been developed rather well. The primary functions of LBS for tourism are usually regarded as being the localization of persons, objects, and places, routing between them, search for objects in proximity such as restaurants, shops, hotels, or points of interest, and information about travelling conditions, such as traffic-related data. Currently, mobile services facilitate the reservation of last-minute trips, rental cars, and hotels; and they provide information about changes and delays of flights and trains, offer guides on restaurants, events, and sightseeing opportunities at the destination (Eriksson 2002; Berger & al. 2003).

Most LBS for tourism have been designed for use in urban settings or along a road network with the purpose of making routing easier in built-up areas, and improving access to different services, such as restaurants, pharmacies, museums, banks or ATMs. The special requirements of services used in nature areas - without a proper road infrastructure and other basic facilities - make it difficult to duplicate the existing service concepts in these areas.

Due to the fact that virtually all tourists travelling to destinations in marginal regions are mainly seeking adventure and excitement provided by nature, the main focus of services supporting this purpose is to enable better accessibility to nature, which denotes guidance, delivery of up-to-date information about the surrounding environment, and safety related services, such as the ability to send emergency messages with accurate position information. In addition to these three aspects, tourism destinations have to face the increasing challenges of sustainable tourism, which not only pertain to the sustainable use of the environment but also to the increased environmental awareness among the tourists. LBS can be considered a novel way of responding to the increased demand for real-time information about the environment. Another possibility for LBS is as a steering tool to protect delicate areas, which checks the volume of visitors and their movement in the area (Dias & al. 2004a). For example, LBS applications can be used to direct tourists away from the most vulnerable areas.

The MINNE project

In an effort to study the integration of technologies to promote new ways of developing sustainable tourism in peripheral areas, the multidisciplinary project "Mobile Environmental Information Systems and Services" (MINNE) was launched in 2002 as a collaboration between researchers from different backgrounds. The general objective of MINNE is to study and develop technologies, methods, models and business solutions for information management utilizing mobile internet technologies. The application domain is mobile sustainable travel and related environmental services. In MINNE, the application domain is approached from the viewpoint of multi-channel adaptive service provisioning and mobile data collection (Antikainen & al. 2004).

The research includes general-level topics such as positioning and location-based information utilization, community services, mobile learning, environmental measurement and sensors, management of information in multi-channel settings and mobile data collection, management and distribution in vehicles and transportation. In the initial phase of the project, the emphasis has been on reviewing, analyzing and evaluating the technological and business possibilities of LBS applications. In a later stage, the main focus of the project will be to develop a generic location-aware environmental information service for use on a smart phone. This service is designed for implementation mainly in botanical gardens, natural parks and other nature tourism destinations.

Tourism industry partners

The project is being carried out in collaboration with some of the most important regional tourist resorts. The resorts represent a set of different types of tourist areas and destinations with different needs and requirements regarding the possible implementation of LBS (Figure 1, Table 1). The research project serves as a framework for investigating the specific type of needs, requirements and feasibility of implementing LBS at each of the resorts, as well as for research into relevant business models and availability and applicability of different types of technologies.

Currently the most active study resort is the Rokua tourism area consisting of a formation of scenically unique eskers, originally moulded during the Ice Age. However, the unique natural environment, mostly comprised of barren forest types, cannot support large numbers of tourists and still remain attractive and in its natural state. The lichen heaths found in the area become damaged easily, exposing the fine-grained esker sand beneath (Hynönen 2004). The need to protect the physical environment

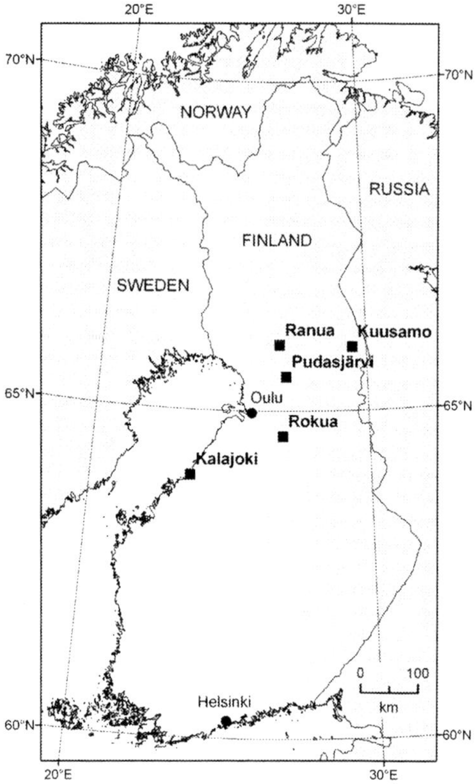

Figure 1. The location of the collaborative tourism resorts.

in the face of a large number of tourists visiting the area, leads to the assessment of new ways of using the route data along with other environmental data in location-based mobile applications that can ensure a more sustainable use of the environment for recreational purposes.

With regard to the potential benefits that LBS may have to offer to tourism services and sustainable tourism, the Rokua tourism area is a case in point. In spite of the current focus on Rokua, however, a goal of the project is to construct general models and concepts for the implementation of LBS in different tourist resorts in Northern Finland.

Table 1. The collaborative tourism resorts and their expected potential for LBS, as classified to four categories: information, guidance, safety and services contributing to environmental protection.

Name	Type	High season	Information	Guidance	Safety	Environmental protection
Kalajoki	coastal resort, large beach	summer	moderate	high	low	low
Kuusamo	downhill skiing center, national park	winter	high	very high	very high	high
Pudasjärvi	downhill skiing center, national park	winter	high	high	high	moderate
Ranua	zoo	summer	very high	moderate	low	low
Rokua	rehabilitation center, unique landscape area	summer	moderate	very high	high	very high

This is possible because, to a large extent, the resorts do not compete for the same segment of the tourism market.

Related projects

Some related research projects aimed at developing LBS for tourism in peripheral natural areas, are or have been in progress. Among the best examples are the Hypergeo and the WebPark projects (Mountain & Raper 2000, 2001), funded by the EU 5th Framework Programme Information Society Technologies (IST), which study the potential of location-based services in tourism and professional use. The Hypergeo established expertise in tourism-related use and the expertise is now being utilized in the current WebPark project. The WebPark project (Mountain & al. 2003; Dias & al. 2004b) is developing personalized, location-aware and timely access to national parks' information systems for mobile devices. They both share many characteristics with the ReGeo project, which aims at developing a decentralized tourist information system based on multimedia and geodata. Test areas include national parks in Austria and the Czech Republic, a landscape park in central Poland, and a nature park in Germany (Almer & al. 2002).

In Finnish Lapland, the Aurora Borealis Testing Lab Programme is studying the integration of new technologies as part of the regional tourism expertise. The focus of testing is on the applications' functionality, usability, and revenue-generation model. The mobile testing programme is the first testing that operates in genuine conditions using real users and concentrating on tourism and experience services. The applications include a service that helps tourists to traverse the countryside along the chosen route, an application package that comprises of a real-time monitoring system of vehicles based on positioning technology and its application for snowmobiles increasing safety and customer satisfaction, and the "Pocket Resort" service offering ski resort visitors a mobile service using current information on the resort's services, events and weather, as well as providing guidance services using location technology maps (A World First... 2003).

In addition to the aforementioned projects, several tourism-related research projects have been launched particularly in the member countries of the European Union with a special emphasis on utilizing the potential of mobile location-based services in the development of tourist industry. In most cases the services are designed for urban settings (Blechschmied & al. 2003; Schmidt-Belz & al. 2003), but there are some projects that seek to use LBS in promoting tourism especially in rural and mountainous areas, including the projects like PARAMOUNT (Löhnert & al. 2001), CRUMPET (Poslad & al. 2001), and TourServ (Schmidt 2001). With the increasing importance of tourism for regional development, the number of projects studying the development of technology and integration of different data sources in tourism is likely to continue growing in the future.

Business models

Due to the failure of mobile services to meet the expectations that many business analysts had of the growth and revenue potential of the telecommunications industry at the turn of the decade, business models pertaining to LBS have been under critical scrutiny. Several approaches to business models for mobile applications and tourism have been assessed in detail (Andersen 2002; Berger & al. 2003; Pääjärvi 2004), and will not be covered here. There are, however, some characteristics that distinguish LBS from regular mobile services. First of all, one of the characteristics of LBS is that its value chain is extremely complicated as it involves a large number of stakeholders. These include network operators, mobile device manufacturers, positioning technology providers, service providers and developers, and consumers (Andersen 2002; Nokia 2004). In the case of a tourism resort operating in a marginal region, the complicated value chain inevitably forms a substantial threshold to investing in LBS, as no actor alone is able to create an end-to-end solution (Pääjärvi 2004).

Even more importantly, while regular mobile services for tourism serve mostly as an additional channel for information dissemination and thereby conform to already well-established business models, location-based applications represent a new and innovative type of service, that require careful consideration by the tourism industry with regard to expectations concerning the earning logic and revenue-generation (Eriksson 2003).

As with any new enabling technology or application area, the actual benefits yielded by it may easily be exaggerated and subject to misconceptions, which may lead to the formulation of unfeasible business models. LBS are no exception to this (Rao & Minakakis 2004). From a practical point of view, the uncertainty as to what kind of services the consumers really want to use, and what they are ready to pay for, are some of the most challenging factors for business model assessment. Some studies concerning the user preferences of LBS specific relevance to tourism have been conducted in the past few years. According to the results, people generally prefer simple and non-intrusive services, where the user maintains control over the information content. Among the most important features are basic services, like maps for orientation purposes, security-related information, and transport information while on the move. This is quite obvious, as the benefits of LBS are most apparent in the transport phase and at the destination, when the traveller may no longer be able to rely on the information he or she has gathered in the planning phase of the trip (Edwardes & al. 2003; Krug & al. 2003; Schmidt-Belz & al. 2003).

Considering that most of the conventional services on the internet are free, it is fairly difficult to justify the potentially high and unpredictable pricing of many LBS for tourism. Although the results of some user surveys indicate that consumers are willing to pay for mobile services, either by pay per use or subscription modes, in most cases it is not feasible to expect the services to generate enough direct cash flow to recoup all the development and maintenance costs, even in the long term. Consequently, LBS should be considered a way to provide added value to the tourism product as a whole. In this respect, the main task is to ensure that the services are developed in a way that they provide true utility to the user, i.e. the value of the service should be higher than the effort made to use the service. A well-designed package of services providing improved access to tourism services, more in-depth information provision and interaction with the environment, and safety, is bound to enhance customer satisfaction and increase demand and use of other tourism services in the region.

Implementation challenges

The successful implementation of LBS must be able to overcome some key challenges, which include the provision of useful and revenue-generating services, smooth integration with different technological systems and high value of data (Kivera 2002). While the new technologies promise benefits and added value, they also raise challenges concerning usability, accessibility over different devices, trustworthiness and interactivity (Werthner & Ricci 2004). Additional challenges relate to the collection and usage of location information, physical quality of devices and the overall market situation.

Technological development

The enabling technology has been evolving rapidly, however this has led to application development from a technology-centric perspective driven by the vendors of technical equipment (Eriksson 2002). Due to the rapid progress in technology the development of common standards has tended to lag behind. As a result, the wide variety of mobile handsets, operating systems, data formats, and limitations in user interfaces, for example, makes it difficult to create generic LBS that could run on any device and be seamlessly connected to the data sources of different providers.

Positioning technology

Although positioning technology involves a smaller number of confusing alternatives than the other technological components of LBS, positioning can in many cases become a constraining factor. The two main options for receiving location information on a mobile device are the satellite-based GPS and positioning based on the mobile network (provided that the device is connected to a mobile network). They have pros and cons and to date there has been no clear indication by the mobile industry as to which technology is preferable for LBS. While GPS enables fairly accurate positioning everywhere in the world, it requires the use of a separate receiver device and does not work indoors or in places where the sky view is obstructed by tall buildings, mountains or dense vegetation. Network-based methods, in turn, require no external devices and thereby a positioning service is readily available to all the devices connected to the network. This positioning method, however, is very imprecise which makes it unsuitable for services requiring location information with an accuracy of at least tens of meters.

Devices fit for field use

Besides adequate technical performance, devices intended to be used for LBS in nature resorts, have to be strong enough to stand substantial variation in temperature, moisture, drop-shocks, vibration, and dust. In addition to these, battery capacity and the luminous intensity of the display affects the usability of mobile handsets in field conditions. Reliability is an important factor especially in areas where the availability of safety-related services have a central role in facilitating more casual wandering in nature. In the current situation, though, most of the mobile devices aimed for the consumer market are not rugged enough to withstand the rough field conditions.

Data sources

The availability of relevant data may vary significantly between different resorts. In the context of LBS for nature tourism, relevant data can be considered as referring basically to up-to-date data about routes (both roads and walking paths), topographic data, information about the most essential tourist services in the area as well as information about various points of interest. Although route and geographic information is usually provided by public organizations, data related to services and sights are generally privately owned and subject to confidentiality restrictions. Furthermore, data usually comes in a variety of different and proprietary formats, which leads to incompatibility problems. Integrating different data sources into a single interface and service is probably one of the most challenging tasks in the development of LBS for any field of industry, and often requires tailor-made data integration instead of the utilization of generic LBS platforms.

Privacy and security

Privacy and security concerns are issues often raised in conjunction with LBS. The current regulations concerning the use of personal positioning information are often considered incomplete and confusing. In terms of personal privacy, much depends on the actual delivery mode of the services, i.e. whether the services are activated by the user ("pull mode") or triggered without a specific request for the services ("push mode") (Easton 2002; Hjelm 2002). Another aspect of privacy and security issues concerns different corporate and organizational data sources that can become exposed to new security risks if incorporated into an LBS system.

Market volume

One of the biggest challenges for developing LBS and adopting them in practice in marginal tourist resorts is the limited volume of the market. LBS are complex systems that require a relatively large investment on product development, data integration and technical infrastructure. With limited revenue prospects, it is difficult to attract big business to invest in LBS for regional tourism industry. For this reason, the development of LBS for tourism in marginalized areas hinges considerably on local efforts and risk investments.

Conclusions: Expected impact on tourism and regional economy

Location-based services are undoubtedly becoming an increasingly important area of interest for different industrial and organizational sectors. The tourism industry is among the first to seek benefits provided by LBS. By providing an ability to navigate more easily and safely in an unfamiliar environment and to look for services and sights of interest, LBS have indeed already proved their potential for providing added value to tourism.

Regardless of the fact that most of the implementations of LBS in tourism to date have been in urban areas, LBS have a significant potential for marginal regions as well. In fact, it is in these regions that the true added value of mobility can be realized. As described earlier in this paper, the significance of LBS for tourism in marginal regions is at least three-fold. Firstly, LBS can be utilized to improve accessibility, information and service provision, and safety, which in itself benefits both the tourists and the tourism resort. Secondly, LBS can be considered as a tool for steering the movement and behaviour of tourists in such a way that the balance between the use of natural environment and its protection is optimized. This is also relevant to the general understanding of the need for more sustainable tourism in terms of both environmental and social aspects. Thirdly, while the content development for LBS for tourism is to a large extent dependent on local activity and investments, the development of LBS may actually help create information intensive jobs in the marginal regions.

The main challenges, however, lie in the usability and reliability of the technology, database interoperability, and the willingness of the tourism industry to invest in a business whose direct revenue potential remains uncertain. The technological challenges are mainly attributable to the rapid and uncoordinated development of mobile handsets and data transfer technologies which have somewhat overshadowed the importance of quality content for mobile services. However, the advances in the technological performance of mobile handsets are an essential enabling factor for LBS.

A more serious problem relates to the availability, interoperability and quality of tourism-related data sources. This essentially calls for the need for agreement on copyright and quality standard issues of data and, particularly, the adoption of common data exchange formats. Due to the numerous efforts needed to define common standards for LBS, the situation is likely to improve significantly in the near future.

Finally, the fact that mobile services in general seldom produce revenue - at least from the content provider's point of view - investing in LBS requires a strong commitment and ability to perceive the value adding role of LBS in the tourism business. On the other hand, a tourism product which embraces the benefits of new technology can be expected to bring opportunities to tourism resorts operating in marginal regions so as to gain a crucial competitive advantage over those resorts sticking only to the core business. LBS should not be expected to provide enough direct cash flow to compensate for all of the implementation costs, but should be seen as a value-adding part of the tourism product, with the added advantage of promoting safe and sustainable tourism.

References

Aichholzer, G., M. Spitzenberger & R. Winkler (2003). Prisma Strategic Guideline 6: eTourism. Austrian Academy of Sciences, Institute of Technology Assessment 3.11.2006. *http://www.prisma-eu.net/deliverables/sg6tourism.pdf*

Almer, A., H. Zeiner, C. Derler, H. Stelzl & P. Luley (2002). National Park Information System. *16th International Conference "Informatics for Environmental Protection"*, September 25-27, 2002, Vienna University of Technology.

Andersen (2002). Digital Content for Global Mobile Services. European Commission Directorate – General Information Society, Final Report 3.11.2006 *ftp://ftp.cordis.europa.eu/pub/econtent/docs/mobilestudy_en.pdf*

Antikainen, H., A. Colpaert, N. Jaako, J. Rusanen, D. Bendas, M. Myllyaho, M. Oivo, P. Kuvaja, J. Similä, K. Marjoniemi, K. Laine & E. Saari (2004). Mobile Environmental Information Systems. *Cybernetics & Systems*, vol. 35: 7–8, 737–751.

A World First! Lapland – the initiation of an experience and tourist service facility with genuine users (2003). *The Rovaniemi Times*, February 2003, 7.

Beaulieu, M. & M. Cooper (2001). *Wireless Internet Applications and Architecture*. Addison-Wesley Professional, Boston.

Berger, S., H. Lehmann & F. Lehner (2003). Location-based Services in the Tourist Industry. *Information Technology & Tourism*, vol. 5: 4, 243–256.

Blechschmied, H., D. Holweg, A. Schilling & U. Jasnoch (2003). Mobile Tourism – New GIS-based Applications at the CeBIT Fair. *Computer Graphik topics*, vol. 15: 1, 17–19.

Butler, R., M. Hall & J. Jenkins. (1998). *Tourism and Recreation in Rural Areas*. John Wiley & Sons, Chichester.

Dias, E., E. Beinat, C. Rhin & H. Scholten (2004a). Location aware ICT in addressing protected areas' goals. *In* Prastacos, P. & M. Murillo (eds.): *Research on computing Science (special edition on e-Environment)*, vol. 11, 291–302. Centre for Computing Research at IPN. Mexico City.

Dias, E., C. Rhin, R. Haller & H. Scholten (2004b). Adding Value and Improving Processes Using Location-Based Services in Protected Areas: The Webpark Experience. *In* Prastacos, P. & M. Murillo (eds.): *Research on computing Science (special edition on e-Environment*, vol. 11, 291–302. Centre for Computing Research at IPN. Mexico City.

Easton, J. (2002). *Going Wireless – Transform Your Business with Mobile Technology*. Harper Business, New York.

Edwardes, A., D. Burghardt & R. Weibel (2003) WebPark: Location Based Services for Species Search in Recration Area. *In Proceedings of 21st International Cartographic Conference*. Durban, South Africa.

Eriksson, I. (2003). Final report of the Working Group "Mobile Services for Tourism" of the project "Working Together for the Future of European Tourism" 3.11.2006. *http://ec.europa.eu/enterprise/services/tourism/tourism-publications/documents/mobileservices.pdf*

Eriksson, O. (2002). Location Based Destination Information for the Mobile Tourist. *In* Wöber K.W., A.J. Frew, M. Hitz (eds.): *Information and Communication Technologies in Tourism*, 255–264. Springer-Verlag, Wien.

Hjelm, J. (2002). *Creating Location Services for the Wireless Web*. John Wiley & Sons, New York.

Hynönen, A. (2004). The usability of the environmental management system in Rokua area. Nordia Geographical Publications. *NGP Yearbook 2003*, Vol. 32:2.

IFN Bulletin No. 17A (2001). *Special Supplement "Tourism and Europe"*. International Friends of Nature, Wien.

Kivera (2002). Value & Applications for Location-Based Services. An Overview of Opportunities & Challenges for Deploying State-of-the-Art Location Technologies 31.10.2004 *http://www.kivera.com/downloads/whitepaper.pdf*

Krug, K., W. Abderhalden & R. Haller (2003). User Needs for Location-Based Services in Protected Areas: Case Study Swiss National Park. *Information Technology & Tourism*, vol. 5: 4, 235–242.

Löhnert, E., E. Wittmann, J. Pielmeier & F. Sayda (2001). PARAMOUNT- Public Safety & Commercial Info-Mobility Applications & Services in the Mountains. *14th International Technical Meeting of the Satellite Division of The Institute of Navigation ION GPS 2001*, September 11-14, Salt Lake City, USA.

Maguire, D. (2001) Mobile Geographic Services Come of Age: ESRI Dives into Wireless Markets. *GeoInformatics*, vol. 4, 6–9.

Mountain, D. & J.F. Raper (2000). Designing geolocation services for next generation mobile phone systems. *Proceedings of the Association of Geographical Information*, Paper W2.4, 1–8.

Mountain, D. & J.F. Raper (2001). Positioning techniques for location-based services (LBS): characteristics and limitations of proposed solutions. *ASLIB Proceedings*, vol. 53: 10, 404–412.

Mountain, D., K. Krug & D. Phan (2003). WebPark: Location-based services for mobile users in protected areas. *GeoInformatics*, vol. 6, 26–29.

Nokia (2004). Location aware applications take off. Nokia white paper. 30.10.2006 *http://nds2.ir.nokia.com/NOKIA_COM_1/About_Nokia/Press/White_Papers/pdf_files/location_wp.pdf*

Poslad, S., H. Laamanen, R. Malaka, A. Nick, P. Buckle & A. Zipf (2001). CRUMPET: Creation of User-Friendly Mobile Services Personalised for Tourism. *In Proceedings of the 3G 2001 – Second Int. Conf. on 3G Mobile Communication Technologies.*

Pääjärvi, M. (2004). Future Mobile Data Services for Tourism. Barriers and Enablers for Adoption of a Mobile Tourist Guide in the Tourism Industry of Norrbotten. Luleå University of Technology, MSc Programmes in Engineering – Department of Business Administration and Social Science / Division of Industrial Marketing and e-Commerce 2004:270 3.11. 2006. *http://epubl.luth.se/1402-1617/2004/270/LTU-EX-04270-SE.pdf*

Rao, B. & L. Minakakis (2004). Assessing the Business Impact of Location Based Services. *Proceedings of the 37th Hawaii International Conference on System Sciences*, January 5-8, 2004.

Saarinen, J. (2003). The Role of tourism in peripheries: tourism, regional development and the issue of social capital in Northern Finland. *In conference proceeding: Perspectives on tourism in Nordic and other peripheral areas*, CD-ROM, University of Umeå, Umeå.

Schmidt, A. (2001). Providing Information on the Slope – Information Access for the Alpine Tourism. TCMC2001, *Telecommunications and Mobile Computing Conference*, Graz, Austria.

Schmidt-Belz, B., A. Zipf, S. Poslad & H. Laamanen (2003). Location-based mobile tourist services – first user experiences. *In* Frew, A. J., M. Hitz & P. O'Connor (eds.): *Information and Communication Technologies in Tourism*, 115–123. Springer Computer Science, Berlin.

Veijalainen, J., K. Virrantaus, J. Markkula & T. Vagan (2001) Developing GIS-Supported Location-Based Services. *In* Ozsu, T., H-J. Schek & K. Tanaka (eds.): *Proceedings of the 2nd International Conference on WEB Information Systems Engineering (WISE 2001)*, 3-6 Dec. 2001, Kyoto. Kyoto University, 423–432.

Werthner, H. & F. Ricci (2004). Electronic Commerce and Tourism. *Communications of the ACM*, vol. 47: 12, 101–105.

Zipf, A. & R. Malaka (2001). Developing location based services for tourism – The service providers view. *In* Sheldon, P. J., K.W. Wöber & D.R. Fesenmaier (eds.): *Information and Communication Technologies in Tourism*, 83–92. 8th International Congress on Tourism and Communications Technologies in Tourism (ENTER 2001), Montreal, Canada. Springer.

From the World of Geography

This column presents talking points and events that have taken place in the world of Geography and geographers. In the current issue we take a look at an excursion to the north.

Excursion on the edge of Europe

Marja Anttonen, Tanja Löytynoja & Katri Suorsa, Department of Geography, University of Oulu

At a few year intervals, the Geographical Society of Northern Finland organises excursions to some geographically interesting places abroad. During the past ten years the excursions have been made to Estonia, Russian Karelia and Romania. The latest trip was realised in September 2006 when a group of 18 people travelled to Finnish Lapland and northern Norway. This time the excursion was planned together with Atlas, the student guild of geographers in the University of Oulu and the subdivision of the society. In this review three of the participants write about the experiences the group had along the way.

The six-day excursion was carried out as a tour. We travelled by a chartered bus from Oulu to Kilpisjärvi, Tromsø, North Cape and Inari (Figure 1). The destinations were chosen on the grounds of general interest and diversity – the city of Tromsø acted as a counterbalance to northern nature, and we took hiking into our program to make our limbs supple after a long time of sitting on the bus. Accommodation varied from a bed and breakfast to cottages and the research station.

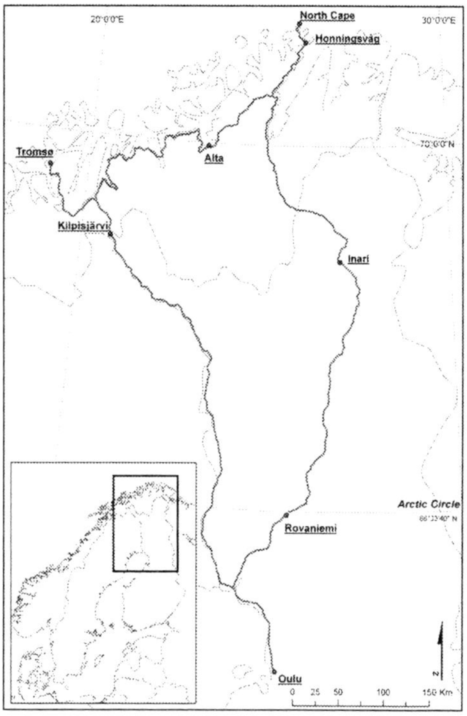

Figure 1. The travel route of the excursion.

Bon voyage!

We left the University of Oulu on Friday, the 15th of September at 2 pm. Everybody was in time and there was excitement in the air. The bus seemed nice, and the same can be said about the participants and the driver. The outward journey to Kilpisjärvi felt surprisingly short – maybe good company, the excitement, a couple of coffee breaks and Matti (the movie) – fooled us to feel that it was only a couple of hours instead of nine. It was dark when were arrived in Kilpisjärvi, so we did not see that much. However, at that time all we needed was a sauna and a bed, and that we got at the Kilpisjärvi Biological Station.

Saturday morning dawned beautiful. The first thing we saw when opening the curtains was Saana – the fjeld we all had heard of but most of us had not seen except in pictures. After the generous breakfast we packed our things and started our climb to Saana. There were 744 steps in the beginning, after that, just rocks and a path that was luckily clearly marked. While walking we had to stop often to take a breath and of course, admiring the view. Other fjelds, like Pikku-Malla, seemed like little hills beside the Saana, whose top is 1,029 metres above the sea level. The whole trip to the top of the Saana and back was eight kilometres. And the trip down was actually more difficult than the way to the top. The reasons were strong headwind, rolling stones and the steps that haunted us (or actually our calves) days after the trip.

After the trip we went back to the Kilpisjärvi Biological Station, had a shower and a good dinner and continued our trip to Norway. Rain started almost in the minute we drove to the Norwegian side of the border, but it did not prevent us from seeing the beauty of the landscapes, which were totally different from the Finnish ones. Flashlights flared in the bus and we stopped quickly to take pictures on the shore of the Arctic Ocean, just as every tourist does.

Tromsø, the Gateway to the Arctic

On the evening we arrived in Tromsø, which seemed like a really beautiful town. The city centre is situated on an island between the continent and another bigger island, Kvaløya, and is surrounded by mountains and sea. Houses are built partly on the mountain slopes, and they seem almost like seabird nest colonies on a cliff. The town has a population of about 63 000 and it is growing in size as people from the surrounding areas move in. Also about 300 Finnish citizens live in Tromsø, mostly working there as nurses or construction workers. Tromsø has the nickname "Gateway to the Arctic" because many arctic expeditions have started there. Also nowadays a route to the Arctic Ocean passes through Tromsø and the sea is strongly present in the city. For example, you can see great ocean liners going under the big bridge right beside the city centre.

During our two-day visit in Tromsø, we, for example, got acquainted with the nightlife of the city which really did not differ from Finnish one, except that smoking is not allowed inside the bars and beer is bit more expensive. On Sunday we walked in the town, drove around the island and visited the polar information centre of Polaria. In Polaria we watched panorama films of Spitsbergen and Antarctic and admired the playful bearded seals, various fish and other creatures (e.g. starfish and sea anemone) in the arctic aquarium.

Next morning we visited the University of Tromsø, the world's northernmost university, established in 1968. Karen Marie Christensen, the Head of the Information department, gave us an introduction about the University and a tour around the campus, which architecture and art give an impressive atmosphere. With about 6 650 students the University is a bit smaller than the University of Oulu. There is no geography department at Tromsø, but other related subjects, like geology, biology and social sciences. It is also possible to study Finnish at the University of Tromsø and we happened to run into one of the teachers along the tour. Among other things she told us about northern Norway's ethnic group of Kvens, who have originally emigrated from northern Finland, and whose language nowadays resembles a mixture of Finnish and Norwegian.

Tromsø seemed like a very lively small town. Maybe it was because of the differences in the altitude, nice-looking colourful old and new houses peacefully side by side, but the town made us, girls from the flat city of Oulu, really excited. Some of us are even dreaming about going back as an exchange students or researchers, but this time we continued our journey towards the north.

Destination 71° 10' 21" N

The trip from Tromsø to North Cape was unbelievable. The landscape was amazing, like in a movie scenery. The roads were mountainous and curving and the slates surrounded the roads. There was not much traffic because the tourism season was over. We stopped for a quick dinner in Alta and continued our trip to the island of Magerøya where North Cape is situated. A route to Magerøya goes through a tunnel that is 6.8 kilometres long and lies 212 metres below the sea level. The road maintenance charge in the tunnel is quite expensive but it is used to cover the cost of building the tunnel.

It was again dark when we arrived in our overnight accommodation in Skipsfjorden, about 20 kilometres from North Cape. The four cottages seemed cosy and warm: there was no electricity, only gas lamps gave light. On Thursday morning, in daylight, we saw a beautiful fiord around us and it was a shame that we did not have time to admire the view longer.

Our next stops were two fishing villages, Kamøyvær and Skarsvåg where we took a short walk. In Skarsvåg we visited Jul og vinterhus and Kirkeporten, which was a geographer's dream – a gate shaped stone formation of slates (Figure 2). In the summertime the midnight sun shines through the "gate", and from Kirkeporten you can also see the Horn of North Cape.

After leaving Kirkeporten we drove towards North Cape, the northernmost point of Europe. On the way we got some foretaste of the extreme weather conditions of this latitude because suddenly it started to snow. In no time, the ground become white and visibility weakened. For a while we were not sure if we can reach North Cape at all, but luckily our bus driver managed to drive us safely there. Besides snowing, it rained, hailed, and even the sun shone during our three-hour visit to North Cape.

The road ended on the cliff which has come to be known as the top of Europe. However, this impressive cliff 307 metres above the sea level is not really the northernmost point of Europe because the nearby cape, Knivskjelodden, extends even farther north (Jacobsen 2000). It is possible to reach this point by hiking, but unfortunately we did not have enough time for it. So, we had to be satisfied with "the touristic extreme point" and visited North Cape Hall, a big stone building partly

Figure 2. Joyful travellers in Kirkeporten. Photo: Tanja Löytynoja 19.9.2006.

constructed inside the cliff. This attraction includes for example a supervideograph, a small historical exhibition, a souvenir shop, a post office, a restaurant and an atmospheric ecumenical chapel. Outside the building is situated the Globe, the famous monument, in front of which millions of tourists have posed for a photograph. We acted equally like all the tourists: we admired the panoramic view over the Arctic Ocean and took a lot of photographs by the monument in spite of the windy weather. At the hall we enjoyed the film about North Cape's nature and bought some souvenirs. There is definitely a chance to get rid of your money at North Cape Hall because of an expensive entrance fee and a great amount of different North Cape products.

Every year over 200 000 tourists visit North Cape. According to Nordkapp booklet (2004), there can be even 60 buses parked in front of North Cape Hall every day during summer time. No wonder that the hall is so big. Within our visit there were five busloads of tourists coming from a cruise ship which was anchored in Honningsvåg, one of the largest cruise ship harbours in Norway.

Back to Finland

The fifth day of our excursion continued with a long drive to Finland. We stopped shortly at Honningsvåg, the main town near North Cape, and again at Trollholmsund

Figure 3. The official group photo in the Artic Circle. Photo: Mika Kärkkäinen 20.9.2006.

where we had a short walk to "the trolls". According to a Sámi legend, these white dolomite deposits are originally trolls that were transformed into stone because they could not find shelter from the sunrise (Nordgård 2004). Despite our good imaginations, we did not manage to see the columns as trolls but with a geological mind, and in the sunset, they were quite impressive. We crossed the Norwegian-Finnish border at nightfall and arrived at our next accommodation, Youth Holiday Centre Vasatokka in Inari just before midnight. Most of us still took a sauna before falling asleep in the nice cottages.

Just like northern Norway, northern Finland is the home of Sámi people. We familiarised ourselves with Sámi culture next morning when we visited in SIIDA, Sámi Museum and Northern Lapland Nature Centre in Inari. There we had a look at the joint exhibition of the museum and nature centre, and watched the multimedia show about northern lights.

The snowfall of the previous day had followed us, and our homeward journey began in the middle of wet snowfall. That is why we had to change our plans and pass sightseeing on the top of Kaunispää, the fjeld near Saariselkä village and skiing centre. Instead, we had a couple of stopovers farther south. For example, travelling from north to south was memorialised by taking some group photographs on the Arctic Circle in Rovaniemi (Figure 3). On Wednesday night the 20th of September, we had come full circle and returned to Oulu. Now it was time to unload the photos and begin to reminisce about our excursion with many experiences (some of the photos can be found on the web-site of our society http://terra.oulu.fi/psms.htm).

References

Jacobsen, J.K.S. (2000). Tourist Perceptions of the Ultimate European Periphery. *In* Brown, F. & D. Hall (eds.): *Tourism in Peripheral Areas: Case Studies*, 74–90. Channel View Publications, Clevedon.

Nordgård, I. P. (2004). Trollholmsund 14.2.2006. *http://www.visitarcticnorway.no/*

Nordkapp – Honningsvåg, the town by the North Cape (2004). To-Foto AS, Harstad.

List of contributors

Antikainen, Harri Department of Geography, P.O. Box 3000, FIN-90014 University of Oulu, Finland. harri.antikainen@oulu.fi

Anttonen, Marja Department of Geography, P.O. Box 3000, FIN-90014 University of Oulu, Finland. marja.anttonen@oulu.fi

Colpaert, Alfred Department of Geography, P.O. Box 111, FIN-80101 University of Joensuu, Finland. alfred.colpaert@joensuu.fi

Forbes, Bruce Arctic Centre, P.O. Box 122, FIN-96101 University of Lapland, Finland. bforbes@ulapland.fi

Heikkinen, Hannu Department of Art Studies and Anthropology, TAIDA, P.O. Box 1000, FIN-90014 University of Oulu, Finland. hannuh@cc.oulu.fi

Heikkinen, Olavi Department of Geography, P.O. Box 3000, FIN-90014 University of Oulu, Finland. olavi.heikkinen@oulu.fi

Karvonen, Jari Department of Information Processing Science, P.O. Box 3000, FIN-90014 University of Oulu, Finland. jari.t.karvonen@oulu.fi

Kumpula, Timo Department of Geography, P.O. Box 111, FIN-80101 University of Joensuu, Finland. timo.kumpula@joensuu.fi

Laine, Kari Thule Institute, P.O. Box 7300, FIN-90014 University of Oulu, Finland. kari.laine@oulu.fi

Luoto, Miska Thule Institute, P.O. Box 7300, FIN-90014 University of Oulu, Finland. miska.luoto@oulu.fi

Löytynoja, Tanja Department of Geography, P.O. Box 3000, FIN-90014 University of Oulu, Finland. tanja.loytynoja@oulu.fi

Myllyaho, Mauri Department of Information Processing Science, P.O. Box 3000, FIN-90014 University of Oulu, Finland. mtmylly@tols16.oulu.fi

Oivo, Markku — Department of Information Processing Science, P.O. Box 3000, FIN-90014 University of Oulu, Finland. markku.oivo@oulu.fi

Robbins, Paul — University of Arizona, Department of Geography and Regional Development, Harvill Building, 437A, Tucson, AZ, USA 85721. robbins@email.arizona.edu

Rusanen, Jarmo — Department of Geography, P.O. Box 3000, FIN-90014 University of Oulu, Finland. jarmo.rusanen@oulu.fi

Similä, Jouni — Department of Information Processing Science, P.O. Box 3000, FIN-90014 University of Oulu, Finland. jouni.simila@oulu.fi

Stammler, Florian — Arctic Centre, P.O. Box 122, FIN-96101 University of Lapland, Finland. florian.stammler@ulapland.fi

Suorsa, Katri — Department of Geography, P.O. Box 3000, FIN-90014 University of Oulu, Finland. katri.suorsa@oulu.fi

Vartiainen Sami — Department of Geography, P.O. Box 3000, FIN-90014 University of Oulu, Finland.